FROM LEARNING THEORY TO CONNECTIONIST THEORY:

Essays in Honor of William K. Estes

Volume 1

Participants at Harvard symposium honoring W. K. Estes (September, 1989). (1) R. D. Luce, (2) C. Krumhansl, (3) R. Falmagne, (4) M. Gluck, (5) J. Kagan, (6) P. Holzman, (7) E. Hilgard, (8) D. LaBerge, (9) G. Bower, (10) D. Medin, (11) J. Young, (12) G. Wolford, (13) G. McKoon, (14) J. Townsend, (15) C. Lee, (16) C. Izawa, (17) P. Suppes, (18) D. Noreen, (19) E. Smith, (20) S. W. Link, (21) R. Day, (22) S. Kosslyn, (23) W. K. Estes, (24) V. Reyna, (25) A. Healy, (26) R. Shiffrin, (27) R. Ratcliff (photograph courtesy of S. W. Link).

FROM
LEARNING THEORY
TO
CONNECTIONIST
THEORY:

Essays in Honor of
William K. Estes

Volume 1

Edited by
ALICE F. HEALY
University of Colorado
STEPHEN M. KOSSLYN
Harvard University
RICHARD M. SHIFFRIN
Indiana University

Psychology Press
Taylor & Francis Group
New York London

First Published by
Lawrence Erlbaum Associates, Inc., Publishers
10 Industrial Avenue
Mahwah, New Jersey 07430

Transferred to Digital Printing 2009 by Psychology Press
270 Madison Ave, New York NY 10016
27 Church Road, Hove, East Sussex, BN3 2FA

Library of Congress Cataloging-in-Publication Data

Essays in honor of William K. Estes / edited by Alice F. Healy,
   Stephen M. Kosslyn, Richard M. Shiffrin.
        p.   cm.
     Includes bibliographical references and indexes.
     Contents: v. 1. From learning theory to connectionist theory — v.
   2. From learning processes to cognitive processes.
     ISBN 0-8058-1097-8 (hard : v. 1). — ISBN 0-8058-0759-4 (hard : v.
   2). — ISBN 0-8058-1098-6 (pbk. : v. 1). — ISBN 0-8058-0760-8 (pbk.
   : v. 2)
     1. Learning, Psychology of—Mathematical models.   2. Cognition.
   I. Estes, William Kaye.   II. Healy, Alice F.   III. Kosslyn, Stephen
   Michael, 1948–   .   IV. Shiffrin, Richard M.
   BF318.E86   1992
   153.1'5—dc20   91-39697
                                                                          CIP

Publisher's Note
The publisher has gone to great lengths to ensure the quality of this
reprint but points out that some imperfections in the original may be apparent.

Contents

Preface

These two volumes are dedicated to the distinguished research career of William K. Estes. The first volume is entitled, "From learning theory to connectionist theory" and emphasizes mathematical psychology; the second volume is entitled, "From learning processes to cognitive processes" and emphasizes cognitive psychology. We will not attempt to recount here the details of Bill Estes' illustrious and productive career, which included training at the University of Minnesota followed by faculty positions at Indiana, Stanford, The Rockefeller, and finally Harvard University. Much information about Bill's career is included in two illuminating autobiographical chapters (Estes, 1982, 1989). The chapters in the present volumes are written by many of Bill's students and colleagues who mark different stages of his career. Undoubtedly, there are many other of his friends, colleagues, "children," "grandchildren," and "step children" who would desire to add their contributions, but space limitations did not make this possible. Bill has admitted that the number of students who have worked with him is so great that "even a list would be formidable" (Estes, 1989, p. 122). Despite the large number of his friends and students, it is clear that Bill has had a profound and lasting impact on each one of us, and we all hold Bill in a special place in our hearts. This deep and sincere affection should be evident from the brief reminiscences or personal comments included in many of the chapters in the present volumes. In addition, to convey a more intimate view of Bill's interactions with his students and friends, we present next some particularly appropriate comments made by one of his students, Robert A. Bjork; these comments were given at a banquet honoring Bill at Harvard University on September 16, 1989.

REFERENCES

Estes, W. K. (1982). *Models of learning, memory, and choice: Selected papers* (pp. 337–364). New York: Praeger.
Estes, W. K. (1989). In G. Lindzey (Ed.), *A history of psychology in autobiography* (Vol. VIII, pp. 94–124). Stanford, CA: Stanford University Press.

A.F.H.
S.M.K.
R.M.S.

William Kaye Estes as
Mentor, Colleague, and Friend

Robert A. Bjork
University of California, Los Angeles

In 1975, at a dinner held at the Meetings of the Psychonomic Society in honor of the 25th anniversary of the publication of Bill Estes's "Towards a statistical theory of learning," I described Bill as a pure accumulator of research interests. He has been able, somehow, to move his research focus over the years from mechanisms of reinforcement and punishment in animals, to human verbal learning and memory, to visual information processing, and to problems of choice and categorization without really abandoning any of those interests along the way. Other speakers at the 1975 dinner stressed his equally remarkable ability to anticipate new developments in a given field.

How are we to understand how one person could be so flexible and innovative while also maintaining a striking continuity in the style and standards that characterize his research? Part of the answer, I think, was provided by Bill himself at that 1975 dinner. David LaBerge had entertained those present with a story about Bill's famous long pauses in one-on-one conversations, saying that he had once got up the courage to ask Bill to explain those long response latencies. He said he outlined what seemed to be the reasonable hypotheses—methodical central processing, rapid processing coupled with excessive editing, rapid processing leading to a conclusion that a given comment or question did not merit a timely reply, and so forth—and asked Bill which was correct. David reported that Bill's response latency to that question had yet to be recorded.

When it was Bill's turn to speak that evening, he commented on the fact that a number of people had referred to his long pauses and that people had also wondered how he had been so able to anticipate important new directions in the field. He said he was prepared to answer both questions, which he did in a singularly clever and gracious fashion. Looking out over the collection of his students, colleagues, and friends in attendance, he said that there was one answer to both questions, "During those pauses, I have been listening to you."

In a very real sense, I *do* think Bill Estes has been listening to us. We, his students, colleagues, and postdocs, have brought our interests to Bill. Those interests have sometimes been triggered by Bill's own work, but sometimes not. He has been willing to listen to us in either case, and he has helped to shape the quality, rigor, and direction of our work. We, in turn, have had the chance to influence Bill Estes. Did Bill, for example, trigger the interest of Doug Medin and Ed Smith in categorization, or was it the other way around? I think the latter was the primary direction of influence. He has done us all the honor of thinking that we may have a good idea—that our new problem may be important and interesting, or that we may have seen a new way to approach an old problem.

Bill has been able to work effectively with all of us, males and females, younger and older, those of us who won't shut up and those of us who are painfully shy, the modelers and the empiricists. He has put his stamp on all of us, but that stamp has nothing to do with telling us what to study. He has never pressured his students to adopt his interests or theories. Saying that one is an Estes student does not have much value as a predictor of one's research domain. His stamp has to do with a commitment to the experimental method, a commitment to rigorous theorizing, and a commitment to scholarship.

Bill's flexibility and ability to listen have kept him young in spirit and in tune with current events. People of his age tend, I have observed, to get together with their cronies at meetings like Psychonomics. No doubt we, too, not that many years from now, will be eating dinner together and grumbling that graduate students are not like they used to be, and that in one way or another the field has gone to hell. Not Bill. He and Kay are surrounded by younger people at such meetings. The "Estes Dinner," which has become a tradition at Psychonomics, has nearly gotten out of hand; getting reservations for a group of that size has become difficult.

In the interests of providing a completely veridical picture, it should be emphasized that Bill is not always the easiest person in the world with whom to work. A serious problem for many of us is that Bill is an impossible model. There is such a thing as having *too* much respect for a person—in the sense that it can impede the natural give and take in the collaborative process. On any of a number of dimensions, it seems hopeless to set one's sights on matching Bill's accomplishments—as a scientist, as a writer, and as a human being. I remember as a graduate student feeling depressed at the ease with which Bill seemed able to improve my laboriously-generated scientific prose. I despaired of ever becoming a good writer. One particularly devastating instance took place when Bill was working over a draft of my dissertation. On one page, where I had labored to clarify a complex point, Bill drew one of his faint pencil lines from the middle of a sentence in the first paragraph—through the entire second paragraph—to the middle of a sentence in the third paragraph, and he replaced everything in between with the single word "provide."

Bill's contributions as an editor, of course, would stand by themselves as an impressive career achievement for most people. In addition to editing books such as the *Handbook of Learning and Cognitive Processes* series, he has edited the *Journal of Comparative and Physiological Psychology, The Psychological Review,* and—at the present time—*Psychological Science*. At the 1975 dinner, David Grant read a series of devastatingly funny excerpts from some of Bill's reviews of manuscripts submitted to the *Journal of Experimental Psychology* while Grant was Editor. They were devastating and funny because such damning comments were stated in such a charitable and gentle fashion (e.g., "In work submitted to this journal, it is customary for there to be an independent variable . . ."). John Swets and myself were recipients of a more recent example of Bill's gentle phrasing: "In the paragraphs following the first one we certainly run into a stretch of rather dry reading." Note that, as stated, Bill does not blame us for that deadly stretch of prose; rather, it is as though he is saying that we have this common problem—that we are all in the same boat. Given that John Swets had drafted that particular section, I, of course, thought Bill's comment was more humorous than John did.

A final unusual aspect of working with Bill that merits mention is that he is impossible to interrupt. Some of his response latencies may set records, but once he is launched on a point or comment he wants to make he has unusual inertia. Even Pat Suppes, in their joint seminars at Stanford, was typically unable to break in once Bill got going. In a famous incident in the Friday-afternoon research talks at Ventura Hall, Karl Pribram did manage to break in, but, as it turned out, at his own peril. A student working with Karl had presented some results that seemed to defy interpretation. Bill started to put forth a possible interpretation by saying, "suppose there are a series of little drawers in the brain . . . ," at which point Karl interjected, "I have never seen any drawers in there." Without missing a beat, Bill said, "They're very small" and continued with his argument.

It is my goal in these remarks to speak not only for myself but for all of us who have been privileged to be Bill's students and, later, his colleagues and friends. I would be remiss if I did not add some final comments about Bill as a person and about his relationship with Kay. Bill has always been there for us, whether that meant never missing a Friday-afternoon seminar, being in his office at Indiana/Stanford/Rockefeller/Harvard when we needed to see him, or when we dropped by from out of town to visit, or being out there in the audience when our paper was scheduled for an early Saturday morning at Psychonomics. He has worried about us, written letters for us, and opened his home for us. Except to Kay, possibly, he may never have uttered an uncharitable word about any of us. In small ways and, sometimes, in very big ways, he has been there when different ones of us needed him.

We have been doubly enriched by being Bill's students, colleagues, and friends because that has meant that we have fallen under Kay's wing as well. It is impossible to reflect on Bill Estes as a person and as a scientist without thinking of Bill and Kay as an inseparable and mutually complementary team. Their warmth, their humor, and their mutual respect have been a joy for all of us. They each in different ways act as though the other is a profound treasure. They are both right.

1 Estes' Statistical Learning Theory: Past, Present, and Future

Patrick Suppes
Stanford University

THE PAST

The direct lineage of statistical learning theory began in 1950 with the publication in *Psychological Review* of Estes' article "Toward a statistical theory of learning." Before saying more about that I recall, however, that there were a number of earlier attempts to develop a quantitative theory of learning which can easily be cited, but I hasten to say that I am not attempting anything like a serious history of the period prior to 1950. I have used the following highly selective but useful procedure of listing the earlier book-length works referred to in the spate of important papers published between 1950 and 1953, which I discuss in a moment. The earlier references referred to (not counting a large number of articles) are Skinner's *Behavior of Organisms,* published in 1938, Hilgard and Marquis' *Conditioning and Learning,* published in 1940, Hull's *Principles of Behavior,* published in 1943, and Wiener's *Cybernetics,* published in 1948. These and other earlier works not mentioned played an important role in developing the theoretical concepts of learning used in statistical learning theory and related theories developed by others shortly thereafter. The basic concepts of particular importance are those of association and reinforcement expanded into the concepts of stimulus, response, and reinforcement, and the processes of stimulus sampling and stimulus conditioning. Other important psychologists who contributed to this development and who were not mentioned on the basis of the principle of selection I used are Guthrie, Thorndike, Thurstone, Tolman, and Watson.

The work of all these psychologists would seem to be almost a necessary preliminary for detailed quantitative and mathematical developments. (Wiener's

1

work, although mathematical in content, is not really in the tradition of psychological research.) The central contribution of statistical learning theory is the use of the psychological concepts of association and reinforcement to develop a genuinely quantitative theory of behavior. When I say *genuinely quantitative,* I have in mind something rather specific. Earlier attempts at quantitative theory, perhaps especially Hull's, did not lead to a theory that was mathematically viable. In Hull's theory it is impossible to make nontrivial derivations leading to new quantitative predictions of behavior. Hull's heart was certainly in the right place and he performed a real service in the kind of efforts he undertook, but it must be said that they were not in any deep sense successful. In contrast, beginning with Estes' 1950 paper, there was a rapid development of ideas that had the same sort of feel about it that the development of ideas and theories have in physics. This is not meant to make any lame comparison between physics and psychology, but rather to make the point that mathematically formulated theory in science, whether it be physics, psychology, or any other discipline, must be set up in such a way that nontrivial quantitative predictions can be made, which can themselves be checked by new experiments or new observations, without some unreasonable number of parameters left to be estimated for each new situation. The many experiments conducted by Estes, his colleagues, and his students over the decade and a half after 1950 attest to the ability to adapt the theory in a fruitful and interesting way to many different experimental configurations.

I turn now to that beginning of the new era in learning that I properly date with Estes' 1950 paper. The opening sentence sets the tone of this new beginning:

> Improved experimental techniques for the study of conditioning and simple discrimination learning enable the present-day investigator to obtain data which are sufficiently orderly and reproducible to support exact quantitative predictions of behavior.

By the end of 1953 a number of important theoretical articles had appeared that set the tone for another decade. I mention especially and in chronological order the two *Psychological Review* articles in 1951 of Bush and Mosteller, which presented in clear and workable mathematical fashion models of simple learning and models of stimulus generalization and discrimination; the 1952 article by George Miller on finite Markov processes; and the 1952 article by Miller and McGill on verbal learning. I end with the 1953 *Psychological Review* article by Estes and Burke on the theory of stimulus variability. These are not the only articles on learning that appeared during this period, nor even the only theoretical ones. But they are by the psychologists who created a new theoretical approach in psychology. To an unusual degree, Estes' 1950 article marks in a very definite way the beginning of this new era.

My own involvement with Bill Estes and learning theory began after 1953. We first began to work together in 1955 when we were both fellows at the Center for the Advanced Study in the Behavioral Sciences at Stanford. It is probably fair to say that I learned more about learning theory in the fall of 1955 in my almost daily interactions with Bill at the Center than at any other time in my life. He brought me quickly from a state of ignorance to being able to talk and think about the subject with some serious understanding. I had previously tried to read the works of Hull mentioned earlier, but had found it fairly hopeless in terms of extension and further development, for the reasons already indicated. Once Bill began explaining to me the details of statistical learning theory I took to it like a duck to water, because I soon became convinced that here was an approach that was both experimentally and mathematically viable.

Over the years I have become involved in other aspects of mathematical psychology, especially the theory of measurement, the theory of decision making, and the learning of school subjects such as elementary mathematics. In the process I have had the opportunity to work with colleagues who have also taught me a lot, particularly Duncan Luce, Dave Krantz, Amos Tversky, Dick Atkinson, and Henri Rouanet. But although many years have passed since Bill and I worked together on a regular basis, I still remember vividly my own intellectual excitement as we began our collaboration in that year at the Center.

More generally in the 10 years following that initial burst of activity from 1950 to 1953, Bill had many good students and much good work was published both by him, his colleagues, and his students. In fact, in my own judgment the decade and a half from 1950 to 1965 will be remembered as one of those periods when a certain new approach to scientific theory construction in psychology had a very rapid and intense development, perhaps the most important in this century.

THE PRESENT

I think it is fair to say that in the 1970s and 1980s much of Bill's own interests shifted from learning theory to problems of memory and related areas of cognition. During this period the new wave of enthusiasm was for cognition rather than for learning. Fortunately, in the last few years we have had a return to a central concern with learning in the recent flourishing of connectionism and the widespread interest in neural networks. It is apparent enough that many of the new theoretical ideas have a lineage that goes back to the developments in learning theory between 1950 and 1965.

I am not going to try to give a survey of these new ideas (see Suppes, 1989), but rather give three applications that arise out of statistical learning theory and related work on stochastic models of learning in the 1950s. These applications are also not meant to be a survey, but reflect my own current work. My purpose in discussing them is to show how the basic ideas of statistical learning theory remain important and have the possibility of application in quite diverse areas of theoretical and practical concern.

My first example concerns the learning of elementary mathematics. The work I report here has been done with Mario Zanotti and will be written up in detail elsewhere. The second example concerns robots that learn and the work I report has been done in collaboration with Colleen Crangle. The third example concerns the learning of Boolean functions, with particular attention to the introduction of hierarchies in such learning. This work has been done with Shuzo Takahashi.

Learning Elementary Mathematics

What I describe under this heading is extensive work at Computer Curriculum Corporation over the past decade on computer-assisted instruction in basic skills, especially elementary mathematics. The fundamental problem is the continual dynamic assessment of mastery on the part of the student. The use of computers for instruction permits deep concern for operational individualization of instruction. This means that each student can be moved forward on an individual basis as he or she masters successive skills or concepts. The course entitled "Mathematics Concepts and Skills" is aimed at supplementary instruction, ranging from kindergarten to the eighth grade. The course is divided into 16 content strands. A strand itself is ordered into a sequence of equivalence classes of exercises of increasing difficulty. An equivalence class is meant to be a collection of exercises of essentially the same difficulty. For example, an equivalence class in the addition strand might contain exercises of adding two three-digit numbers with no carry from one digit to another. There are also strands concerned with measurement, geometry, word problems, and so on. I do not attempt to give here a systematic description of the content of the course, which in a broad sense overlaps extensively with any standard textbook series. The features of individualization and the organization of a given strand into an ordered sequence of equivalence classes are aspects that differ radically from the structure of textbooks. The computer-assisted instruction session presented to a student on a given day includes exercises from a selection of the strands appropriate to the student's grade level. In general the exercises are selected across the strands on a random basis according to a curriculum distribution, which is itself adjusted to be a convex mixture with a purely subjective distribution depending upon the individual student's strengths or weaknesses. The curriculum probability distribution is itself something not to be found in the general theory of curriculum or in textbooks, but is essential to the operational aspects of the kind of course I am describing. I do not here, however, go into the details of either the curriculum distribution or the individual student distribution, nor do I discuss the equally important problems of contingent tutoring when the student makes a mistake, or how a student is initially placed in the course.

The detailed use of learning theory occurs in monitoring and evaluating when a student passes a mastery criterion for leaving a given equivalence class of exercises to move on to a more difficult concept or skill. Here we have been able to apply detailed ideas that go back to the framework of statistical learning theory introduced in Bill Estes' 1950 paper, but with an application he probably never originally had in mind: to the daily activity of students in schools in many parts of the country. Classical and simple ideas about mastery say that all that is needed is a test. On the basis of the relative frequency of correct answers in a given sample a decision is made as to whether the student shows mastery. But the first thing that one finds in the detailed analysis of data is that when a student is exposed to a class of exercises, even with previous instruction on the concepts involved, the student will show improvement in performance. The problem is to decide when the student has satisfied a reasonable criterion of mastery.

The emphasis here is on the data analysis of the actual learning, not on the setting of a normative, criterion of mastery. For this purpose some standard notation is introduced. Although in almost all of the exercises concerned, the variety of wrong student responses is large, I move at once to characterize responses abstractly as either correct or incorrect. With this restriction in mind I use the following familiar notation:

$A_{0,n}$ = event of incorrect response on trial n,

$A_{1,n}$ = event of correct response on trial n,

$\quad x_n$ = possible sequence of correct and incorrect responses from Trial 1 to n inclusive,

$\quad q_n$ = $P(A_{0,n})$, the mean probability of an error on Trial n,

$\quad q = q_1$,

$q_{x,n}$ = $P(A_{0,n}|x_{n-1})$.

Also, $\underline{A}_0$ and $\underline{A}_1$ are the corresponding random variables.

The learning model that Zanotti and I have applied to the situation described is one that very much fits into the family of models developed by Estes, Bush, Mosteller, and the rest of us working in the 1950s. In addition to q, there are two parameters to the model. One is a uniform learning parameter α that acts constantly on each trial, because the student is always told the correct answer; and the second is a parameter w, which assumes a special role when an error is made. This is one way to formulate the matter. A rather more interesting way perhaps is to put it in terms of the linear learning model arising from statistical learning theory analyzed very thoroughly in Estes and Suppes (1959, 1959a, 1959b). The parameter α corresonds to $1 - \theta$ in the linear model derived from statistical learning theory. The linear learning model derived from statistical learning theory puts all of the weight as such on the reinforcement. The generalization considered here is straightforward and can be found in earlier studies as well. In terms of the two parameters α and w we may write the basic assumptions of the model in terms of the following two equations:

$$P(A_{0,n+1}|A_{0,n}, x_{n-1}) = (1 - w)\alpha P(A_{0,n}|x_{n-1}) + \alpha w, \tag{1}$$

$$P(A_{0,n+1}|A_{1,n}, x_{n-1}) = (1 - w)\alpha P(A_{0,n}|x_{n-1}). \tag{2}$$

It is then easy to prove by familiar methods that in terms of random variables:

$$E(\underline{A}_{0,n+1}) = \alpha^n q. \tag{3}$$

$$\text{Var } (\underline{A}_{0,n+1}) = \alpha^n q(1 - \alpha^n q). \tag{4}$$

Equation (3) just expresses in terms of random variables the familiar mean learning curve, which also holds in the learning model of statistical learning theory, but written in terms of θ rather than in terms of α. A variety of methods are available for estimating the three parameters of the model, namely, the parameters q, α, and w, but I do not go into such methods here.

In Figs. 1.1–1.4 I show four mean learning curves for third- and fourth-grade exercises in addition and multiplication of whole numbers and addition of fractions. The grade placement of each class of exercises is shown in the figure title,

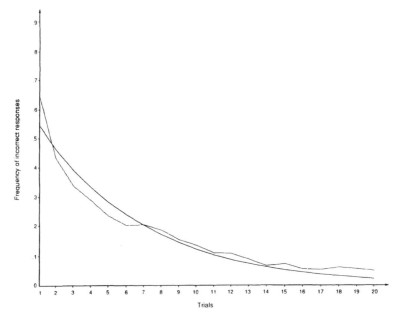

FIG. 1.1.   Addition at grade-level 3.65, sample size = 612, $\hat{q}$ = 0.545, $\hat{\alpha}$ = 0.847.

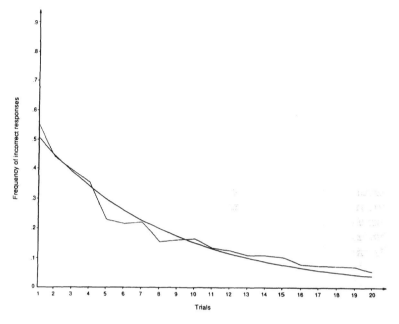

FIG. 1.2. Multiplication exercises at grade level 3.65, sample size = 487, $\hat{q}$ = 0.508, $\hat{\alpha}$ = 0.875.

for example, 3.65 for multiplication. But this does not mean that all the students doing these exercises were in the third grade, for with the individualization possible, students can be from several different chronological grade levels. The sample sizes on which the mean curves are based are all large, ranging from 406 to 616. The students do not come from one school and certainly are not in any well-defined experimental condition. On the other hand, all of the students were working in a computer laboratory run by a proctor in an elementary school so there was supervision of a general sort of the work by the students, especially in terms of schedule and general attention to task. In these mean learning curves and the sequential data presented later, the students who responded correctly to the first four exercises have been deleted from the sample size and from the data, because in terms of the mastery criterion used, students who did the first four exercises correctly in a given class were immediately moved to the next class in that strand. No further deletions in the data were made. For each figure the estimated initial probability q of an error and the estimated learning parameter α are given.

The data and theoretical curves shown in Figs. 1.1–1.4 represent four from a sample of several hundred, all of which show the same general characteristics;

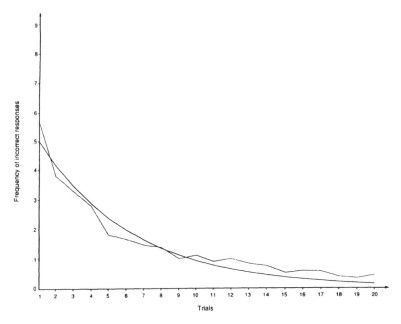

FIG. 1.3.   Multiplication exercises at grade level 3.95, sample size = 406, $\hat{q}$ = .500, $\hat{\alpha}$ = 0.831.

namely, very rapid improvement in probability of a correct response as practice continues from the first trial onward. In most cases the student will have at least one intervening trial between exercises from a given class. So, for example, between two fraction exercises there might well intervene several different exercises, one a word problem, another a decimal problem, and so on. Also, it is probably true for all of the students that they had had some exposure by their classroom teacher to the concepts used in solving the exercises, but, as is quite familiar from decades of data on elementary-school mathematics, students show clear improvement in correctness of response with practice. In other words, learning continues long after formal instruction is first given. The most dramatic example of an improvement is in Fig. 1.4. This is not unexpected, because understanding and manipulation of fractions are among the most difficult concepts for elementary-school students to master in the standard curriculum.

In Tables 1.1–1.4, data from the same four classes of exercises are analyzed in terms of the more demanding requirement on the learning model to fit the sequential data. In the present case we looked at the first four responses with, as already indicated, the data for students with four correct responses deleted. This left a joint distribution of 15 cells, and in the case of the sequential data, the

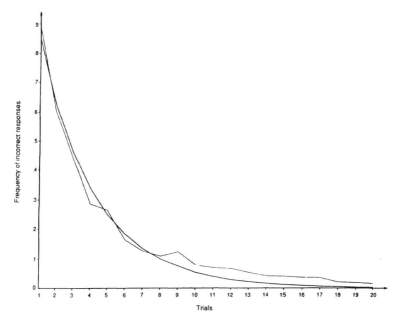

FIG. 1.4.   Fraction exercises at grade level 4.40, sample size = 616, $\hat{q}$ = 0.849, $\hat{\alpha}$ = 0.736.

TABLE 1.1
Observed and Theoretically Expected Distribution for the First Four Responses to Addition
Exercises at Level 3.65.
Sample Size - 612, DF = 11, $\hat{\omega}$ = 0.188, $\hat{q}$ = 0.406, $\hat{\alpha}$ = 0.750, $\chi2$ = 14.39

Cell	Observed	Expected
0000	25	20.6
0001	36	35.0
0010	19	21.1
0011	75	70.6
0100	13	15.6
0101	34	39.3
0110	31	25.5
0111	163	151.6
1000	12	12.7
1001	19	27.3
1010	19	17.3
1011	60	80.0
1100	18	14.6
1101	49	48.3
1110	39	32.6

TABLE 1.2
Multiplication Exercises at Level 3.65.
Sample Size - 488, DF = 11, $\hat{\omega}$ = 0.125, $\hat{q}$ = 0.375, $\hat{\alpha}$ = 0.875, $\chi 2$ = 4.44

Cell	Observed	Expected
0000	17	17.9
0001	25	24.2
0010	20	18.9
0011	40	41.0
0100	19	16.1
0101	33	31.0
0110	30	25.1
0111	87	83.1
1000	13	14.7
1001	25	25.9
1010	16	20.8
1011	59	61.8
1100	15	18.7
1101	46	48.6
1110	43	40.2

theoretical computations involve the parameter w as well as α. For learning theorists the stringency of the test to fit such sequential data with only three parameters is well known. The data in each of the tables have 11 degrees of freedom because of the three parameters estimated from the data. The largest X^2 is for the addition exercises, but even here the X^2 is not significant at the 0.10 level. It is to my mind surprising that the data fit as well as they do.

It is to be noticed that the values of q and α are not the same for the mean learning curve and the joint distributions. This is because we fit each case

TABLE 1.3
Multiplication Exercises at Level 3.95.
Sample Size - 406, DF = 11, $\hat{\omega}$ = 0.188, $\hat{q}$ = 0.281, $\hat{\alpha}$ = 0.781, $\chi 2$ = 9.43

Cell	Observed	Expected
0000	12	9.4
0001	12	16.0
0010	9	10.5
0011	29	36.3
0100	10	8.6
0101	23	22.3
0110	21	15.6
0111	115	103.4
1000	8	8.2
1001	17	18.1
1010	13	12.4
1011	54	62.6
1100	8	11.5
1101	43	41.3
1110	32	29.7

TABLE 1.4
Fraction Exercises at Level 4.40.
Sample Size - 616, DF = 11, $\hat{\omega}$ = 0.313, $\hat{q}$ = 0.813, $\hat{\alpha}$ = 0.688, χ^2 = 8.87

Cell	Observed	Expected
0000	76	73.5
0001	96	89.8
0010	34	38.8
0011	121	126.0
0100	21	21.7
0101	41	40.5
0110	18	21.1
0111	141	136.5
1000	6	7.7
1001	12	11.5
1010	9	5.5
1011	16	23.8
1100	5	4.2
1101	13	9.9
1110	7	5.5

directly and the best estimates were not precisely the same for the two different statistics, as is familiar in learning data. It is also apparent that if we used the same values for both mean learning curves and the sequential data, the fits would still be reasonably good in all four classes exhibited in Figs. 1.1–1.4 and Tables 1.1–1.4.

The main point of my presentation of these data is to show the viability of the learning models that originate in statistical learning theory and related work in the 1950s to nonexperimental school situations. I argue that the examples given show that much deeper application of mathematical and quantitative learning concepts and theories can be applied directly to school learning. This applicability of a very direct kind contrasts quite markedly to any attempt to apply in the same quantitative fashion learning ideas to be found in the earlier work of Hull or Skinner, for example, cited at the beginning of the previous section.

Robots that Learn

In this example of the application of statistical learning theory a more radical attitude is taken toward the use of the theory, because it is not an application directed at the analysis of data from human or animal performance, but rather uses the theory as the basis of a built-in mechanism to smooth out and make robust the performance of a robot that is learning a new task. The idea of such instructable robots, on which I have now worked for a number of years with Colleen Crangle and others (Crangle & Suppes, 1987, 1990; Maas & Suppes, 1985; Suppes & Crangle, 1988) is organized around two leading ideas. First, robots should be able to learn from instructions given in ordinary, relatively vague English, just as we expect children and apprentices to do so in learning a new task. Second, quite apart from the vagueness of the English, even the conceptualization of the task is often relatively vague and qualitative in character. The use of precise coordinates is not easily available to the instructor and consequently probability distributions provide a method of robustly adjusting to the nature of the task—probability distributions, I should emphasize, that are dynamically changing as a consequence of learning.

The approach we use to these matters follows directly from work of my own on learning models for a continuum (Suppes, 1959a, 1959b), which very much falls within the framework of statistical learning theory. The leading idea is that on any trial the robot has a smoothing distribution $\kappa_{\mu,\sigma}(x)$, which earlier I called a smearing distribution. (As should be evident, μ is the mean of the distribution and σ^2 its variance.) The idea is that this is a distribution that is smeared around the conceptually ideal response. The probability of a response lying between a and b is $\int_a^b \kappa_{\mu,\sigma}(x)dx$. Notice that I am implicitly assuming, and now do so explicitly, that the applications are confined to one dimension. In fact, for the present discussion I restrict myself to the beta distribution on the open interval $(0,1)$

$$\beta(x) = \frac{\Gamma(\alpha + \beta)}{\Gamma(\alpha)\Gamma(\beta)} x^{\alpha-1}(1 - x)^{\beta-1}$$

where, as is familiar, the distribution requires that $\alpha,\beta > 0$. Moreover the mean μ and the variance σ^2 are:

$$\mu = \frac{\alpha}{\alpha + \beta}, \qquad \sigma^2 = \frac{\alpha\beta}{(\alpha + \beta)^2(\alpha + \beta + 1)}.$$

Note that the uniform distribution is a special case of the beta distribution when $\alpha = \beta = 1$. In our present work Crangle and I have restricted ourselves to one dimension although there is considerable interest in the two-dimensional applications. But the learning–theoretic ideas become more complicated rather rapidly. A typical one-dimensional task that we may think of for purposes of illustration is a robot opening a door. Its problem is to position itself along an interval parallel to the door when closed. Within that framework we think of three kinds of feedback or reinforcement to the robot: positional feedback, accuracy feedback, and congratulatory feedback, where each of these different kinds of feedback are expressed in ordinary qualitative language. In the case of positional feedback, we have three range constants indicating how much the means should be moved. The constants are qualitatively thought of as large, medium, and small, with the large parameter being 2, the medium 1, and the small 0.5. (The exact value of these parameters is not important.) In this framework we have then the relations between feedback and learning shown in Table 1.5.

Notice that all of the positional feedback commands are not given. It is understood that for each of those to the left, there is a corresponding one to the

TABLE 1.5
Relation Between Feedback and Learning

Feedback/Reinforcement	Learning
Much further left!	$\mu_{n+1} = \mu_n - 2\sigma$
Right just a little!	$\mu_{n+1} = \mu_n + 0.5\sigma$
Move to the left!	$\mu_{n+1} = \mu_n - \sigma$
Be more careful!	$\mu_{n+1} = \sigma_n^2$
No need to be so cautious!	$\mu_{n+1} = \sqrt{\sigma_n}$
Just right!	$\mu_{n+1} = r_n$
	$\sigma_{n+1}^2 = \sigma_n^2$

right, and vice versa. In the case of the positional feedback the variance stays the same, and only the mean is changed. In the case of accuracy feedback, shown in the second part of Table 1.5, only the variance is changed, not the mean. The algebraic expression of the change in variance reflects the fact that we are restricting ourselves in the present example to the open interval (0, 1). Finally, congratulatory feedback is to indicate that the distribution being used is satisfactory. Its location and accuracy is about right and so the mean should be the last response, that is $\mu_{n+1} = r_n$, and the variance stays the same. In order to reduce the number of figures, in Fig. 1.5 I show at each step only the distribution on the basis of which the response is made. The response is shown by a heavy black vertical line and then the verbal feedback that follows use of this distribution to the right of the graph. The graphs are to be read in sequence starting at the top and moving to the right from the upper left-hand corner. I should note that the initial instruction not shown here is "Go to the door," which is also repeated at the beginning of each new trial.

As the graphs are meant to show, with appropriate reinforcement and further trials, the robot ends up with a reasonable distribution for the task of opening the door. It might seem that we should not go to all this trouble of repeated trials and the use of English, when one could just program in from the beginning the correct distribution. This has been very much the attitude in much of the work in computer science on robots, but a little reflection on the wide world of practical activities will indicate that it is a very limited approach to performing most tasks. Complicated problems requiring sophisticated motor and perceptual skills are really never handled by the explicit introduction of precise parameters. Apprentices learn from watching and getting instruction of a general sort from a master craftsman. A youngster learning to play the piano or learning tennis is not told

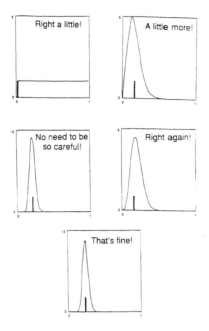

FIG. 1.5.   Illustrations of the changes in probability distribution by an instructable robot learning to open a door.

how to set parameters but is given practice and repeated qualitative instruction on ways to improve the performance. I am not suggesting by these remarks that we have reached a very sophisticated level in the instruction of robots, but just that the road that lies ahead for the kind of work outlined here seems to be conceptually the right approach to a great many tasks we will expect robots to perform in the future. The central point I want to make is that the way we have thought about learning in the past seems to me very appropriate for thinking about learning in the future as far as robots are concerned. In this simple example, for instance, I have certainly bypassed all of the problems of laying out the exact cognitive and perceptual capabilities of the robots to whom the theory is supposed to apply. When one turns to actual examples, it is clear that the cognitive and perceptual capabilities will be severely limited but can in many cases be adequate to specific tasks. It is my own conjecture that the kind of ideas initiated by Bill Estes and others in the 1950s will have a useful, and in many cases central, role to play in the training of robots for specific tasks, not merely in the distant future but rather soon.

Learning Boolean Functions

In this third example I turn to purely theoretical issues, but ones that have been the focus of a number of papers in computer science in the last few years, especially by Valiant (1984, 1985). Much of the work of Valiant and others—I mention especially the recent long article by Blumer, Ehrenfeucht, Haussler, and Warmuth (1989)—has been on learnability of various classes of concepts, including especially Boolean classes. The thrust of this theoretical work has essentially been to show under what conditions concepts can be learned in polynomial time or space. These general polynomial results are important, but I would like to focus here on reporting joint work with Shuzo Takahashi in which we are much more concerned with detailed results of the kind characteristic of learning theory as applied to human behavior. Secondly, the work of Valiant has emphasized the importance of considering only restricted Boolean functions that have some special syntactic form, for example, disjunctive normal form. Although I state some general results that Takahashi and I have obtained, I mainly concentrate on examples, in order to show the power of imposing a learning hierarchy whenever pertinent conditions obtain.

For the analysis of learning Boolean functions, it is very natural to go back for the learning process to one of Bill Estes' most important papers, the 1959 one entitled "Component and pattern models with Markovian interpretations." I especially have in mind Bill's introduction of the pattern model in this paper. The assumptions about the pattern model close to his that we use are these: First, one stimulus pattern is presented on each trial; second, initially all patterns are unconditioned; third, conditioning occurs with Probability 1 to the correct response on each trial. This last assumption is a specialization of the general formulation Estes uses in which conditioning occurs with Probability c. In this application, intended for machine learning, it is appropriate to set $c = 1$. Fourth, there are two responses, 1 for the pattern being an instance of the concept, and 0 for its not so being. Finally, for the initial discussion I make a sampling assumption that is much more special than any assumed in Bill's 1959 paper, but one that is useful for the purposes of clarifying the role of a hierarchy in the learning of Boolean functions. This is the strong assumption that there is a complete sampling of patterns without replacement.

The intuitive idea of the hierarchy is easily introduced, and the reasons for it as well. Suppose we are interested in learning an arbitrary Boolean function of n variables. Then there are 2^n patterns. Without introducing a hierarchy there is, in the general case, no natural way to avoid sampling all of the patterns if we want to learn the concept completely. Even when we want to learn the concept incompletely, but with reasonable confidence of a correct response, if the patterns are distributed more or less uniformly, then learning by sampling each pattern is not feasible—not feasible in the technical sense that it is exponential in n.

On the special learning assumptions introduced, and I emphasize particularly the assumption that patterns are sampled without replacement, it is easy to prove some general results. As a matter of notation let

N_t = number of trials for nonhierarchical learning of term t,
H_t = number of trials for hierarchical learning of t,
V_t = number of distinct variables in t,
S_t = set of function symbols in t,
$A(f)$ = number of arguments in function t.

We may then easily show that

$$N_t = 2^{V_t}$$
$$H_t = \sum_{f \in S_t} 2^{A(f)}.$$

Here are some simple examples:

For $t = f(x,y), N_t = 2^2 = 4$ and $H_t = 2^2 = 4$.
For $t = f(g(x_1,x_2),h(x_1,x_2,x_3))$, $N_t = 2^3 = 8$ and $H_t = 2^2 + 2^2 + 2^3 = 16$.
But for $t = f(g(x_1,x_2,x_4), g(x_2,x_3,x_4))$, $N_t = 2^4 = 16$ and $H_t = 2^2 + 2^3 = 12$.

Note that in each of these examples, f, g, and h are arbitrary Boolean functions. The exact form of the function does not matter. For instance, in the first example f could be Boolean intersection, Boolean disjunction, or Boolean exclusive or. Here is the best sort of hierarchical example:

$$t = h(g(f(x_1, x_2), f(x_3, x_4)), g(f(x_5, x_6), f(x_7, x_8)))$$
$$N_t = 2^8 = 256$$
$$H_t = 3 \times 2^2 = 12$$
$$= 4 \log_2 V_t$$
$$= 4 \log_2 8 = 4 \times 3 = 12.$$

On the other hand, here is the worst sort of hierarchical example:

$$t = h(g(f_1(x_1, x_2), f_2(x_1, x_2)), g_2(f_3(x_1, x_2), f_4(x_1, x_2)))$$
$$N_t = 2^2 = 4$$
$$H_t = 7 \times 2^2 = 28.$$

Introducing one more matter of notation, let $F_t = |S_t|$ be the number of function symbols in t. We may then state the following theorems that are easily proved.

Theorem 1. *For terms with only binary functions, if $V_1 \geq 4$ and $V_t \geq F_t$, then $H_t \leq N_t$.*

Notice that Theorem 1 establishes a general inequality for hierarchical versus nonhierarchical formulations with restriction to binary functions. Terms so restricted we call *binary terms*.

Theorem 2. *For binary terms, if $V_t \geq 4$ and $F_t/V_t \leq r$ then $H_t/N_t \leq r$.*

More than a decade ago (Suppes, 1977) I was emphasizing the importance of hierarchies for efficiency in learning, but at that time I did not appreciate how great the gain might be. I end by describing one simple example that is quite dramatic. Let the number of individual variables be 1024—obviously I pick this number to have a nice power of 2 but the exact number is of no importance. Then the number of patterns is 2^{1024}, which on the basis of individual sampling would take more time to complete than the universe has yet experienced. The number of trials, although finite, is unthinkably large. However, I now introduce the following three assumptions. First, let the hierarchy be of the form of the best example given above but such that each occurrence of the binary function f is now a distinct function f_i. Second, let there be learner control of stimulus sampling, so that the learner can sample exactly the pattern desired; and third, let the processing be parallel. Then the trials to learn an arbitrary Boolean function expressible in this particular hierarchy by binary functions is just 40. This is a reduction from 2^{1024} of more than 300 orders of magnitude! More generally, for n the number of variables, the trials to learn for this best hierarchical example with parallel processing is $4 \log_2 n$.

The proof in the general case is quite simple. The four patterns for parallel processing at the initial stage and recursively then for each next level of the hierarchy can be shown as follows:

$x_1 x_2$	$x_3 x_4$	. . . .	$x_{n-1} x_n$
11	11	. . . .	11
10	10	. . . .	10
01	01	. . . .	01
00	00	. . . .	00

Second, it is easy to show that the number of learning levels in the hierarchy is simply $\log_2 n$. This last example with $n = 1024$ is admittedly a very special case, but, even in a very special case, to obtain a reduction of 300 orders of magnitude in the number of learning trials is something not easy to find in any domain of knowledge under any circumstances whatsoever. We are all intuitively convinced of the importance of hierarchies in the organization of large-scale learning in the schools, but it is very desirable to have a dramatic artificial example to show their power.

THE FUTURE

It is obvious enough that I think statistical learning theory will continue to flourish in the future and have a place of importance. Over the past generation we have raised a lot of cognitive psychologists who know very little about learning of a serious sort, but the new generation is properly learning all about connectionism and neural nets. A natural question is whether the pattern model, as just used in the discussion of learning Boolean functions, will be replaced by neural nets. There is an important distinction to be made here—a distinction that is critical in understanding why statistical learning theory will continue to have a significant future. This distinction is that between methods of implementing learning, and analysis in terms of best results possible. In describing the pattern model I have not at all tried to describe how it would be implemented in hardware or software of either the biological or electronic kind. It is easy enough to show, and should be apparent, that the learning results for Boolean functions just discussed cannot be improved by any new network configuration. In fact, it would undoubtedly be the case that under many kinds of network analysis the learning rate would not be so rapid as that shown for the pattern model. Granted, I mean the pattern model with the strong assumption that the conditioning parameter $c = 1$. This example of course was meant to be artificial, but the same kind of very strong hierarchical results would obtain under a broad range of sampling assumptions about the patterns. What is said there about parallel processing of patterns is critical to efficient learning of a biological as well as electronic kind. The most obvious example is the human vision system, although I am not pretending that a simple analysis in terms of Boolean functions will suffice to provide a detailed theory.

Parallel processing, it seems to me, is the important concept to add to Estes' classical pattern model. This addition will represent an important modification for the future. In contrast, the use of neural networks is not an extension, or in contradiction to, the pattern model, but represents a lower level of abstraction, what I would call a schematic level of implementation. I hasten to add that I strongly favor this lower level of abstraction, for a full account of learning needs a theory of the biological or electronic processing of both external stimuli and internal computations.

On the other hand, as much as such an extended theory is desirable, we can easily be misled into understanding the difficulty of developing it or the usefulness of simplifying abstractions like the pattern model. A pluralism of levels of abstraction and of corresponding models is, in my judgment, a permanent feature of any science of complex phenomena. It is naive and mistaken to think we shall find the one true complete theory of learning based on accurate details of how neurons and their connections actually work. Many different levels of theorizing will continue to be of value in many different domains. There is every reason to think that the kinds of applications of statistical learning theory described in the previous section will have a robust future.

More generally, the classical concepts of association, generalization, and discrimination will be extended, but it seems likely that these basic concepts will continue to play the role in the psychology of learning that the concepts of classical mechanics such as force, mass, and acceleration have played for 200 years of physics. It is not that physics has not developed many new concepts and theories, it is, rather, that once fundamental concepts are put in some reasonable mathematical framework and are recognized as having great generality, they do not disappear. Such will be the future of the fundamental ideas of statistical learning theory.

ACKNOWLEDGMENTS

I am grateful to Duncan Luce for a number of helpful comments on an earlier draft.

REFERENCES

Blumer, A., Ehrenfeucht, A., Haussler, D., & Warmuth, M. K. (1989). Learnability and the Vapnik-Chervonenkis dimension. *Journal of the Association for Computing Machinery, 36*, 929–965.

Bush, R. R., & Mosteller, F. (1951a). A mathematical model for simple learning. *Psychological Review, 58*, 313–323.

Bush, R. R., & Mosteller, F. (1951b). A model for stimulus generalization and discrimination. *Psychological Review, 58*, 413–423.

Crangle, C., & Suppes, P. (1987). Context-fixing semantics for instructable robots. *International Journal of Man-Machine Studies, 27*, 371–400.

Crangle, C., & Suppes, P. (1990). Introduction dialogues: Teaching new skills to a robot. *Proceedings of the NASA Conference on Space Telerobotics, January 31–February 2, 1989*, pp. 91–101.

Estes, W. K. (1950). Toward a statistical theory of learning. *Psychological Review, 57*, 94–107.

Estes, W. K. (1959). Component and pattern models with Markovian interpretations. In R. R. Bush & W. K. Estes (Eds.), *Studies in mathematical learning theory* (pp. 9–52). Stanford, CA: Stanford University Press.

Estes, W. K., & Burke, C. J. (1953). A theory of stimulus variability in learning. *Psychological Review, 60*, 276–286.

Hilgard, E. R., & Marquis, D. G. (1940). *Conditioning and Learning*. New York: Appleton-Century.

Hull, C. L. (1943). *Principles of Behavior*. New York: Appleton-Century.

Maas, R., & Suppes, P. (1985). Natural-language interface for an instructable robot. *International Journal of Man-Machine Studies, 22*, 215–240.

Miller, G. A. (1952). Finite Markov processes in psychology. *Psychological Review, 17*, 149–167.

Miller, G. A., & McGill, W. J. (1952). A statistical description of verbal learning. *Psychometrika, 17*, 369–396.

Skinner, B. F. (1938). *The behavior of organisms*. New York: Appleton-Century.

Suppes, P. (1959a). A linear model for a continuum of responses. In R. R. Bush & W. K. Estes (Eds.), *Studies in mathematical learning theory* (pp. 400–414). Stanford, CA: Stanford University Press.

Suppes, P. (1959b). Stimulus sampling theory for a continuum of responses. In K. J. Arrow, S. Karlin, & P. Suppes (Eds.), *Mathematical methods in the social sciences* (pp. 348–365). Stanford, CA: Stanford University Press.

Suppes, P. (1977). Learning theory for probabilistic automata and register machines, with applications to educational research. In H. Spada & W. F. Kempf (Eds.), *Structural models of thinking and learning, Proceedings of the 7th ISPN-Symposium on Formalized Theories of Thinking and Learning and their Implications for Science Instruction* (pp. 57–79). Bern, Switzerland: Hans Huber.

Suppes, P. (1989). Current directions in mathematical learning theory. In E. E. Roskam (Ed.), *Mathematical psychology in progress* (pp. 3–28). Berlin: Springer-Verlag.

Suppes, P., & Crangle, C. (1988). Context-fixing semantics for the language of action. In J. Dancy, J. M. E. Moravcsik, & C. C. W. Taylor (Eds.), *Human agency: Language, duty and value* (pp. 47–76, 288–290). Stanford, CA: Stanford University Press.

Valiant, L. G. (1984). A theory of the learnable. *Communic. ACM, 27,* 1134–1142.

Valiant, L. G. (1985). Learning disjunctions of conjunctions. In *Proceedings of the 9th International Conference on Artificial Intelligence,* vol 1 (pp. 560–566). San Mateo, CA: Morgan Kaufmann.

Wiener, N. (1948). *Cybernetics.* New York: Wiley.

2 Choosing Between Uncertain Options: A Reprise to the Estes Scanning Model

Gordon Bower
Evan Heit
Stanford University

Throughout his career, Bill Estes has worked on refining theories of how rewards and punishments operate to shape and control organisms' behavior. From his earlier training with B. F. Skinner, Estes acquired a firm commitment to studying the way incentives and reinforcement schedules control behavior in lower organisms; and his later concern with human learning led him into careful study of the research papers of Edward L. Thorndike. Thorndike was an extremely important figure in learning theory in the 1920–1945 period and Estes was a student in the later years. Thorndike's proposed "Law of Effect" was to serve as the focal point around which revolved most studies of reinforcement and learning (Thorndike, 1931).

Thorndike's basic proposal was that trial-and-error learning, of what response to make to what stimulus, only occurs if the subject receives a reward (or "satisfying state of affairs") soon after the correct response. The satisfying effect was said to strengthen the associative bond formed between the stimulus and the response that brought about that effect. That proposal led to considerable controversy, as a number of psychologists of that era (such as Guthrie, Hull, and Tolman) believed that (a) learning of, or memory for, S-R connections required only attention to, and rehearsal of, their contiguous occurrence, and that (b) rewards and punishment acted not by affecting memory-storage but rather by influencing the organism's motivation to convert memories into performance. Specifically, the alternative theorists believed that organisms learned by merely attending to events, but they might not exhibit their learning until the proper motivational and incentive conditions were operating.

For various historical reasons, the empirical battle between these two views was waged within the confines of the animal laboratory, more specifically over

21

the question of whether rats could learn their way around complex mazes even though no food, water, or other biological reinforcer was provided during their initial explorations. This question spawned the scores of latent-learning experiments, which yielded a conflicting, inconsistent morass of unresolved outcomes that nearly became the donnybrook of this style of doing learning theory (see MacCorquodale & Meehl, 1954). The latent learning, rat-in-a-maze experiment proved to be simply too variable and complex to provide conclusive evidence on the point at issue.

Estes changed the test venue of this issue to the human learning laboratory. For this purpose he adapted and modified three different experimental arrangements, all of which had the same analysis within his stimulus sampling theory. The three arrangements were:

1. *Thorndike's task:* To each of *K* single stimuli, human subjects make one of *N* different responses (e.g., *K* letters as stimuli paired with *N* button presses as responses), with different rewarding outcomes or penalties contingent on which response was made to each stimulus. This arrangement is also called "trial-and-error" learning (see Keller, Cole, Burke, & Estes, 1965).

2. *Two-armed bandit task:* To a single stimulus presented over several hundreds of trials, the subject chooses one of two (or more) responses, and receives a reward or penalty selected according to a probability distribution contingent upon the selected response (see Estes, 1962).

3. *Pair comparison task:* Presented on each trial with a pair of stimuli (e.g., A vs. B, C vs. D), the subject chooses one and receives a reward or penalty from a probability distribution contingent on the alternative selected. Feedback information may or may not be given about the reward outcome scheduled for the unchosen response. Tests usually involve new combinations of the alternatives (e.g., A vs. C, D vs. B) (see Estes, 1966).

In a remarkable series of papers in the 1960s (Estes 1962, 1966, 1969; Keller et al., 1965), Estes sharply defined the alternative conceptions of how incentive conditions might operate in controlling learning and performance in both humans and animals. He and his associates marshaled a series of ingenious experiments whose results forcefully argued for the "informational" view (today we would call it the "cognitive" view) of how incentive conditions influence learning. The strong alternative to Thorndike's Law of Effect was that rewarding outcomes simply supply "information" that subjects can learn, regarding what outcomes follow given responses in given situations (see the review in Bower & Hilgard, 1981, pp. 37–44). In animal experiments, it was difficult to get the animal to attend to the critical events (to be learned) without some motive or reward involved. But with verbally proficient humans, one can easily direct attention to to-be-learned events and, at the same time, separate the information-value of a reward from its "satisfying" effect when received. In several clever experiments, Estes arranged conditions in which human subjects received information without "satisfaction," by learning what reward they would have obtained had they chosen the other alternative on the current trial. For example, in an experimental task involving memory overload (Estes, 1969), subjects had to correctly predict which one of two reward values they would receive on a given trial. The subjects learned to prefer the alternative with the higher expected value, even though the values displayed on each trial were selected by the experimenter so that subjects never actually received any reward for that alternative. Subjects' later choices clearly reflected this incentive information despite their never having received any rewards for that learning. The range of results, devastating to Thorndike's position, is reviewed in Estes (1969).

In explaining these results, Estes assumed that rewards direct subjects' attention to contiguous S-R events, and that stimulus-response-outcome connections become interassociated purely by contiguity. We may represent this learning by a schema (see Ross & Bower, 1981) that records the sequential occurrence of mental tokens of the stimulus, then the response, then the outcome (S, R, O). Thus, reinstating the situational stimulus may retrieve the triplet from memory, activating a representation of the response as well as the incentive outcome expected for the response.

This schema theory has to solve the problem of how to translate such S-R-O memories into actions or behavioral choices. Estes (1962, 1969) solved this problem by postulating that stimuli will lead to action only in conjunction with positive feedback from an expected reward. Although Estes formulated the theory so that it was applicable to animals performing operant responses to obtain rewards and/or avoid punishments, he applied it mainly to the behavior of humans in risky choice situations. For application to such decision situations, Estes formulated his theory of a response scanning process by which anticipated consequences influence the behaviors selected. Next we turn to Estes' specific formulation of the scanning model and detail some of its successes. Then we describe some experimental results that are problematic for the scanning model. Finally, we present an alternative model, based on expected utility, that can explain these problematic results as well as the earlier studies accounted for by the scanning model.

Estes' Scanning Model

Estes (1966) was able to capture his ideas about magnitude of reward in mathematical form in what he called the *scanning model.* He formulated the model as follows:

> The gist of the theory can be expressed in two principal assumptions: (a) In any
> situation in which different responses are followed by different rewards with fixed

probabilities, a S's tendency to expect a given reward O_j following any particular response A_i is assumed to be learned according to the same process that governs acquisition of overt responses in simple probability learning (or response members of items in ordinary paired-associate learning). (b) Upon presentation of the stimulus at the beginning of each trial, S is considered to scan (implicitly) the set of available response alternatives, generating for each alternative a prediction of the reward that will be received if the response is made and finally making the response which he predicts will yield the largest reward. (p. 276.)

Estes applied this model to a number of situations, of which we describe two.

The first application is to a pair comparison study (Estes, 1966) in which subjects choose between elements of three pairs of nonsense syllables (denoted here by letters), receiving the points (convertible to money) for choices as indicated: Pair 1 (A-3 vs. E-1), Pair 2 (B-3 vs. D-2), and Pair 3 (C-2 vs. F-1). On each trial, the subject was informed of the point value only for the item selected. Subjects were trained until they consistently chose for each pair the alternative with the larger payoff. Then subjects received a series of no-feedback test choices with the other 12 combinations of stimulus pairs. These tests yielded the choice proportions shown in Table 2.1. The predictions alongside each observed proportion were derived from the scanning model using a second set of data, shown in Table 2.2.

The Table 2.2 data were obtained by asking subjects after learning to indicate how many points each alternative was worth when it was chosen. Table 2.2 lists the group's probability of saying that any given syllable was worth 1, 2, or 3 points.

Generally, subjects learned the correct value, but there was some error in mixing up the 1- and 2-point syllables. This poor learning is especially evident for the least-favored alternative in each pair, for which subjects had fewer opportunities to learn their correct value (e.g., compare C vs. D).

To predict the choice proportions in Table 2.1, for any pair i, j, the scanning process is best described using the following array:

	Reward		
Stimulus	1	2	3
i	a	b	c
j	d	e	f

In this array, each entry is the probability that a subject believes that choice of the row stimulus will lead on a given trial to the reward in that column. In each row, the probabilities will add to 1. The probability that on any given trial when alternative A_i is compared to A_j the scanning process will terminate with choice of alternative i over j is then given by

$$p_{ij} = \frac{c(d + e) + bd}{1 - (ad + be + cf)}. \tag{1}$$

TABLE 2.1
Choice Proportions (Row over Column) on Test Trials[a]

	B	C	D	E	F
A	54 (52; 54)	98 (98; 88)	98 (99; 97)	—	100 (100; 99)
B	—	90 (93; 87)	—	98 (97; 98)	100 (99; 99)
C	—	—	81 (81; 81)	85 (85; 86)	—
D	—	—	—	62 (61; 61)	78 (85; 80)
E	—	—	—	—	60 (76; 72)

[a]In parentheses beside each observed value are the predicted value, first, for the scanning model, and second, for the expected utility model.
Note. From "Transfer of verbal discriminations based on differential reword magnitudes" by W. K. Estes, 1966, Journal of Experimental Psychology, 72, p. 278. Reprinted by permission.

In Equation 1, the numerator is the probability that the subject predicts a larger reward for i than for j on this trial, and the denominator is one minus the likelihood of predicting a tie in outcomes. (Note that the lower case letters a-f designating reward probabilities are not meant to correspond to the upper case letters A-F designating stimulus alternatives.) A critical assumption of the model is that if tied predictions arise, then the subject scans over the alternatives again, generating new independent outcome predictions, continuing thus until the process generates a favored alternative for choice.

For each pair in Table 2.1, one can substitute the appropriate probabilities from Table 2.2 into Equation 1 to calculate the probability of choosing the row over the column alternative. For example, in the D versus E test, using the entries in the D and E rows of Table 2.2, the calculations are:

$$p_{DE} = \frac{.100(.575 + .325) + .475(.575)}{1 - [(.425)(.575) + (.475)(.325) + (.100)(.100)]}$$

$$p_{DE} = .61.$$

TABLE 2.2
Proportions of Reward Predictions on Single-Stimulus Trials

Stimulus	Its Payoff	Alternative's Payoff	1	2	3	Mean
A	3	1	0	.125	.875	2.875
B	3	2	.050	.075	.975	2.825
C	2	1	.050	.825	.125	2.175
D	2	3	.425	.475	.100	1.675
E	1	3	.575	.325	.100	1.525
F	1	2	.800	.200	0	1.200

Note. From "Transfer of verbal discriminations based on differential reword magnitudes" by W. K. Estes, 1966, Journal of Experimental Psychology, 72, p. 279. Reprinted by permission.

This is the proportion predicted for this pair in Table 2.1. Other entries are calculated similarly.

The entries in Table 2.1 show several interesting features. First, people's choices are very sensitive to the reward values. Second, subjects are more accurate in remembering the value of the alternative preferred during training. Although C and D objectively have the same value, subjects were more likely to believe that C was worth 2 points because it had been chosen more during training; consequently, in the C versus D test, C was chosen 91% of the time, and the model's choice proportion reflects these beliefs about point values. Finally, the scanning model predicts the observed choice proportions with remarkable accuracy, and it does so without estimating any parameters. Rather, it merely needs to have the independently-obtained reward-probability estimates given in Table 2.2 along with the prediction formula in Equation 1. The results provide a resounding success for the scanning model.

Two-Armed Bandit Experiments

Another testing ground for Estes's scanning model was the so-called *two-armed bandit* experiments. In such studies the subject receives a series of several hundred trials choosing between the same two alternatives, A_1 and A_2, and receiving different payoffs with different probabilities. In the paradigmatic case, if subjects choose A_1, they receive a payoff of w cents with probably π_1 and lose x cents with probability $1 - \pi_1$; if they choose A_2, they receive a payoff of y cents with probability π_2 and lose z cents with probability $1 - \pi_2$. In Estes' view, this experiment contains two probability-learning tasks, viz. learning the probability of winning or losing after each of the two responses. Assuming that these embedded probability-learning processes proceed trial by trial according to the linear operator changes of stimulus sampling theory, subjects should arrive asymptotically to expect each outcome with its actual probability, that is, expect w cents after A_1 with probability π_1, and y cents after A_2 with probability π_2. Given these outcome expectations, the scanning model then suggests how to calculate asymptotic choice proportions.

To illustrate these predictions, we apply the model to several experiments involving symmetric payoffs and losses. That is, the amounts won were the same for the two responses ($w = y$) as were the amounts lost ($x = z$). The main variable was the probability of winning following the two responses. For these cases, the predictions of the scanning model are best derived by use of the following array:

Possible Outcomes

Pair	A_1	A_2	Probability	Choice
1.	w	w	$\pi_1 \pi_2$	rescan
2.	w	$-x$	$\pi_1(1 - \pi_2)$	A_1
3.	$-x$	w	$(1 - \pi_1)\pi_2$	A_2
4.	$-x$	$-x$	$(1 - \pi_1)(1 - \pi_2)$	rescan

TABLE 2.3
Asymptotic Probabilities of A_1 Responses Observed and Predicted in Twelve Conditions Studied
by Atkinson, by Siegel, and by Friedman, Padilla, and Gelfand

					$P(A_1)$	
Experiment	Group	π_1	π_2	Observed	Predicted Scanning	Predicted Expected Utility
Atkinson	1	.60	.50	.60	.60	.61
(1962)	2	.70	.50	.69	.70	.71
	3	.80	.50	.83	.80	.80
Siegel	1	.75	.25	.93	.90	.91
(1961)	2	.70	.30	.85	.85	.86
	3	.65	.35	.75	.77	.80
Friedman,	1	.80	.80	.47	.50	.50
Padilla, and	2	.80	.50	.81	.80	.80
Gelfand	3	.80	.20	.94	.94	.94
(1964)	4	.50	.50	.48	.50	.50
	5	.50	.20	.82	.80	.80
	6	.20	.20	.47	.50	.50

The subject is presumed to consider each response in turn, predicting a reward
(of w cents) or a loss ($-x$). The third column cites the probabilities that these four
possible pairs of predictions will arise on any particular asymptotic trial after
extensive training. The probability that the scanning process terminates with an
A_1 response is given by the probability of the second relative to the third row.
This is

$$p_{1,2} = \frac{\pi_1(1 - \pi_2)}{\pi_1(1 - \pi_2) + \pi_2(1 - \pi_1)} \qquad (2)$$

$$= \frac{\phi_1}{\phi_1 + \phi_2},$$

where $\phi_i = \pi_i/(1-\pi_i)$. One may also note that if the payoff for A_1 is larger than
for A_2 (i.e., $w > y$), then the first row of scanned predictions as shown would be
added to that favoring A_1; similarly, if the amount lost for A_1 is less than for A_2
(i.e., $x > z$), the fourth row as shown would be added to that favoring A_1.

Some asymptotic choice probabilities from three experiments using symmetric payoffs and losses are shown in Table 2.3. In all these experiments, the
subjects won 5¢ or lost 5¢ for both responses, so only the payoff probabilities
were varied across groups. These values of π_1 and π_2 are listed in Table 2.3;
when substituted into Equation 2, they yield the predicted choice proportions in
the sixth column. Several observations are warranted. Whenever π_2 equals .50,
then $p_{1,2}$ is equal to π_1; this prediction is well supported by the results of Table

TABLE 2.4
Asymptotic Probabilities of the Three Responses in a Symmetric Payoff Experiment
by Siegel and Andrews (1961)

			$P(A_i)$		
Group	Response	π_i	Observed	Predicted Scanning	Predicted Expected Utility
1	1	.65	.78	.81	.81
	2	.25	.12	.14	.13
	3	.10	.10	.05	.06
2	1	.70	.876	.870	.860
	2	.15	.058	.065	.070
	3	.15	.066	.065	.070
3	1	.75	.89	.90	.89
	2	.20	.08	.08	.07
	3	.05	.03	.02	.04

2.3. Whenever $\pi_1 = \pi_2$, then the predicted choice probability is always .50. For fixed π_1, choice of A_1 decreases the greater the payoff probability for response A_2. These aspects of the data are well captured by the scanning model. Indeed, the model predicts these data extremely well; the fact that its predictions are parameter-free makes it even more impressive. The predictions in the final column, labeled "expected utility," are for an alternative theory that we consider later.

As a final illustration of the success of the scanning model, consider its predictions for a three-alternative choice experiment conducted by Sydney Siegel and Julie Andrews in 1961 with symmetric wins and losses (of 5¢) and with the payoff probabilities as shown in Table 2.4. The predictions, shown in Table 2.4, were derived by a simple extension of the scanning process that yields a three-response version of Equation 2, adding $\phi_3 = \pi_3/(1 - \pi_3)$ to the denominator. Again the results provide striking confirmation of the predictions of the scanning model. Several similarly accurate predictions of the scanning model are reported in Estes's papers (1962, 1969).

Readers may well wonder how the scanning model deals with the standard probability-learning situation in which there are no payoffs but only information about whether event E_1 or E_2 occurred. The scanning model assumes that in this case subjects are not motivated to continue rescanning if they generate tied predictions for responses A_1 and A_2; rather, they simply guess either response with probability .50. If so, then the probability of an A_1 response at asymptote should be

$$p_1 = \pi_1(1 - \pi_2) + .5[\pi_1\pi_2 + (1-\pi_1)(1-\pi_2)] = .5 + .5(\pi_1 - \pi_2).$$

In the conventional noncontingent reinforcement case, we have $\pi_1 = 1 - \pi_2 = \pi$, so that p_1 will equal π, which is the familiar case of probability matching. There is, of course, abundant evidence that subjects probability match in such "information only" situations. The scanning model tells us in addition what happens when incentives are introduced, providing motivation for the subject to come to a better-than-guessing decision on most trials. An advantage of the scanning model is its convenient interpretation of the probability-matching result in all those experiments involving no payoffs.

Problematic Results for the Scanning Model

Although the scanning model was remarkably successful in explaining a number of results, it has encountered difficulties in several domains. We review just two of these problematic areas, as they lead naturally into the alternative model that we propose later.

One source of difficulty arises when the payoffs in the two-armed bandit experiments are asymmetric, differing for the two alternatives. We discuss two examples of such experiments. One unpublished experiment by Bower and Michael Lloyd (in 1963) had the format shown in Table 2.5: for Groups 1 and 2, response A_1 won or lost 5¢, and response A_2 won or lost 1¢; the probability of a win following A_1 and A_2 was .65 and .35, respectively, for Group 1, and .75 and .25, respectively, for Group 2. For Groups 3 and 4, the reward magnitudes were reversed: response A_1 won or lost 1¢ and response A_2 won or lost 5¢, with the win probability following A_1 set at .65 for Group 3 and at .75 for Group 4. Ten

TABLE 2.5
Probability of A_1 Response in Different Payoff Conditions[a]

Condition	Payoffs	Proportions	$P(A_1)$ Observed	Predicted Scanning	Predicted Expected Utility
1	$A_1 \pm 5$ $A_2 \pm 1$	$\pi_1 = .65$ $\pi_2 = .35$	.80	.65	.79
2	$A_1 \pm 5$ $A_2 \pm 1$	$\pi_1 = .75$ $\pi_2 = .25$	.92	.75	.90
3	$A_1 \pm 1$ $A_2 \pm 5$	$\pi_1 = .65$ $\pi_2 = .35$	.77	.65	.79
4	$A_1 \pm 1$ $A_2 \pm 5$	$\pi_1 = .75$ $\pi_2 = .25$	.89	.75	.90

[a]From Bower and Lloyd (1963).

subjects were run in each condition for 300 trials, yielding the asymptotic A_1 choice probabilities shown under "Observed" in Table 2.5.

Applying the scanning model to these four conditions, we note that in Groups 1 and 2 the A_1 response should be selected whenever one predicts a win for it (with probability π_1); similarly, in Groups 3 and 4, A_2 would be selected whenever one predicts a win on A_2; so the predicted A_1 probability in these cases would be one minus the probability of a win on A_2. These predictions are shown in the next-to-last column of Table 2.5. (The final column shows predictions from the expected utility model to be presented later.) Choice proportions were sensitive to changes in the payoff probabilities, but in this asymmetric case the results (and the predictions) were nearly the same whether the big payoff and penalty were placed on response A_1 or on A_2. That is, observed proportions for Groups 1 and 3 were nearly equal as were those for Groups 2 and 4. However, the data are systematically higher than the predictions of the scanning model; therefore, these results provide some disconfirming evidence against that theory.[1]

This result is not unique; a similar outcome was reported in an experiment by Myers and Suydam (1964) who used wins and losses of $\pm 4\cent$ for response A_1 and $\pm 1\cent$ for A_2; the win probabilities were $\pi_1 = 1 - \pi_2 = .60$ for their Group 1, and $\pi_1 = 1 - \pi_2 = .80$ for their Group 2. Estes' scanning model predicts asymptotic A_1 proportions of .60 and .80 for these two groups. As before, the observed values, at .70 and .94, were systematically above the predictions of the scanning model.

Beyond this difficulty with data from asymmetric payoffs, a serious problem with the simple formulation of the scanning model is that it treats the magnitude of payoffs or penalties in only an ordinal sense; that is, the reward magnitudes enter ordinally only in the "comparison" phase as the model arrives at its decision.[2] Thus, the model predicts the same results for any set of rewards and losses that preserve the same ordering of outcomes following the various re-

---

[1]Curiously, the scanning model can predict these data more closely if we alter the assumptions to suppose that whenever subjects predict a loss on both responses, then they rescan over new predictions. With this modification, the scanning model then predicts $p(A_1)$ of .84 for Groups 1 and 3, and .92 for Groups 2 and 4. For the Myers & Suydam (1964) experiment, the revised model predicts values of .81 and .95 for the two experimental groups (vs. observed values of .70 and .94).

[2]However, the scanning model may be generalized to allow for repeated scans in the case of asymmetrical payoffs. In these cases, when choosing between very different reward magnitudes, subjects might be described as considering the alternatives more carefully, by scanning more times than do subjects in experiments involving symmetric payoffs. If a choice requires k consecutive scanning outcomes favoring the same alternative, then the probability of choosing i over j becomes $\phi_i^k/(\phi_i^k + \phi_j^k)$, where k is a free parameter corresponding to how many consecutive scanning outcomes each subject requires to make a final choice (Estes, 1962). Although fitting this extra parameter would improve the scanning model's ability to account for the data, this *ad hoc* explanation still has several shortcomings. For example, a repetitive scanning process in itself is still only sensitive to the ordinal properties of rewards.

sponses. Thus, the model expects subjects to behave similarly whether they are winning and losing a few pennies or a few million dollars. We submit that even in the experimental laboratory adults are likely to be more than ordinally-sensitive to the magnitude of wins and losses. Our model proposed below is sensitive to these magnitudes.

The Horse-Race Problem

A second difficulty for the scanning model arises when it is applied to competitive races such as horse races (or runners or cars) and political elections between candidates. Whereas Estes (1976a, 1976b) successfully applied the scanning model to some race data, a basic weakness of the model was pointed out by Neely (1982). In predicting the winner in a race between candidates i and j, the scanning model takes into account only the numbers of past wins and losses of each candidate considered separately. Thus, in the simplest version of the model, a candidate who has won π_i percent of his or her past races will be selected over one who has won π_j percent of his or her past races with a probability of $\phi_i / (\phi_i + \phi_j)$.

Neely (1982) pointed out that this exclusive focus on wins and losses suppresses or ignores information about which opponents the wins and losses were against. In most competitive races and games, a win against a "strong" opponent suggests greater strength of the winner than does a win against a "weak" opponent. In ranking U.S. college football or basketball teams, for example, sports writers consider not only a team's won–loss record but also how "tough" (or strong) their opponents have been. A political newcomer who beats a popular incumbent is considered to be a strongly attractive vote getter. For similar reasons, losing to a strong opponent does not imply nearly so much weakness as does losing to a weak opponent.

Neely sought to test these intuitions in experiments involving repeated, hypothetical election races between several politicians. The experiment first set up one candidate as strong (denoted S) by having him beat another, weaker candidate (denoted W) in 80% of their many, repeated head-to-head elections. Two other, middle-level candidates (M1 and M2) ran against each other repeatedly, winning and losing equally often. In the second phase of the experiment, a novel candidate was run repeatedly against S, the earlier strong candidate; we label him NS for the "Novel Strong" candidate. Another novel candidate was run repeatedly against W, the earlier weak candidate; we label him NW for the "Novel Weak" candidate. A novel, middle-level candidate, NM, was also introduced and run against an earlier midlevel candidate, and each of them won and lost 50% of their races.

The interesting feature of Neely's design is that the three novel candidates introduced in Phase 2, NS,NW, and NM, each won 50% of their races, so they are equally strong if only their win–loss records are considered. Yet candidate

NS was shown by the race pairings in Phase 2 to be equally strong as candidate S; candidate NW was shown to be equally weak as candidate W; and candidate NM was shown to be equal to M's strength, whatever that may be. The question of interest is whether these inferred strengths of candidates NS, NM, and NW will be thusly ordered and reflected in later choice probabilities, or whether they will all be equal due to their equated probabilities of winning in Phase 2.

Neely assessed these strengths by later test choice-trials in which each candidate was placed in competition against every other one (with no feedback provided on such test trials). The results showed that the intuitions were correct: Novel candidates acquired more strength the stronger was the candidate against whom they won 50% of their races. In the most critical comparison, the choice between the Novel Strong and Novel Weak candidates yielded a preference of .67 for the NS candidate. This significantly exceeds the .50 level predicted by the scanning model. Moreover, in nearly all the relevant comparisons, candidate NS proved to be a more attractive candidate than candidate NW. For example, in runoffs against candidate NM, subjects predicted NS would win 54% whereas they predicted NW would win only 37% of the elections. Neely replicated the main results in a second experiment, so the conclusions are fairly reliable.

All the data are consistent with the idea that subjects use winning against specific opponents as clues to the attractiveness or vote-getting strength of a given candidate, and that they use the event of candidate i beating j as a basis for estimating the relative strengths of the two candidates. Moreover, candidate i will be equally strong as candidate j if they beat each other half the time in their races.

The Expected Utility Model

The Neely study and the intuitions behind it led to our developing an alternative to the scanning model. We call it the *expected utility model* in deference to the classical rational theory of choice. We apply it to those data reviewed previously that are problematic for the scanning model as well as those reviewed earlier that supported it.

Our theory is basically the old-style expected utility model in which the strength or attractiveness of a given response or choice alternative depends in a direct way upon the history of payoffs and penalties that have followed that response. In applying this theory based on response-strengths, we wish to make a major distinction between two general behavioral situations. We call these the *absolute* versus *comparative* situations, distinguished by the type of contingencies and feedback the subject receives on each trial.

Absolute Situations

In absolute situations, on each trial, the subject chooses one among several alternatives and is given (or informed of) the reward or penalty for the response selected. In some cases the subjects are told in addition what payoff they would have received had they chosen the other alternative on this trial (Keller et al., 1965). We assume here that the responses can be scaled in their strength or attractiveness; we let $s_i(n)$ index the strength of response alternative A_i at the beginning of trial n of some learning experiment. The learning assumption of the model is that the strength of a given alternative that receives feedback of outcome o_i on trial n will be changed according to Equation 3:

$$s_i(n + 1) = s_i(n) + \theta(u_i(n) - s_i(n)) \tag{3}$$

In Equation 3, $u_i(n)$ is the utility corresponding to the outcome, $o_i(n)$, associated with A_i on trial n. The u_i's can be positive or negative and will be measured on an interval scale (i.e., will be unique up to a linear transformation). *Theta* is the learning rate parameter (between 0 and 1) reflecting the fractional change in $s_i(n)$ brought about by experiencing outcome o_i on trial n. A second assumption is that the strength of responses for which no feedback is provided on a given trial will not change, that is, $s_i(n + 1) = s_i(n)$.

Imagine applying Equation 3 to a K-choice situation in which alternative A_i leads to outcome o_j with probability π_{ij}. By weighting Equation 3 according to the probability of the payoffs, we can derive that the expected value s_i on trial n + 1 (given full information) will be

$$s_i(n + 1) = s_i(n) + \theta(\bar{u}_i - s_i(n)).$$

where $\bar{u}_i = \Sigma_j \pi_{ij} u_{ij}$ is the expected utility of the possible outcomes o_j for response A_i. The solution to this difference equation is

$$s_i(n) = \bar{u}_i - (\bar{u}_i - s_i(1))(1 - \theta)^{n-1}. \tag{4}$$

Equation 4 describes the familiar exponential growth curve, going from an initial value on Trial 1 (usually zero) to the average utility for alternative A_i.

To relate these response strengths to choice behaviors, we assume here that the s_i's are first converted to a Luce-like strength measure, defined as $v_i = e^{\alpha s_i}$. Then the probability of the subject selecting response A_i out of a set 1, 2, 3, . . . , K is given by Luce's choice axiom, viz.

$$p_i = \frac{v_i}{\Sigma_j v_j}. \tag{5}$$

The advantage of using Luce's axiom here is that it permits us to interrelate choices from among differing sets of alternatives. The conversion using the exponential function and α parameter has the additional advantage of being able to account for a range of choice behaviors, from responding according to the ratios among response strengths up to always selecting the alternative with the greatest strength.

Having stated the theory for absolute situations, we now apply Equations 4 and 5 to several experiments of this type considered earlier.

Two-Armed Bandit Experiments. First, let us consider the two- or three-armed bandit experiments of the type described in Table 2.3. In those experiments, each alternative A_i led to a win of 5¢ or loss of 5¢ with probability π_i and $1 - \pi_i$, respectively. The expected utilities as asymptote for the two responses would be

$$\bar{u}_1 = \pi_1 u(5) + (1 - \pi_1)u(-5)$$
$$\bar{u}_2 = \pi_2 u(5) + (1 - \pi_2)u(-5).$$

Substituting these into Equation 5, we obtain the probability of choosing A_1 over A_2, viz.,

$$\begin{aligned} p_{1,2} &= \frac{1}{1 + e^{-\alpha(s_1 - s_2)}} \\ &= \frac{1}{1 + e^{-\alpha(\bar{u}_1 - \bar{u}_2)}} \\ &= \frac{1}{1 + e^{-b(\pi_1 - \pi_2)}}, \end{aligned} \tag{6}$$

where $b = \alpha[u(5) - u(-5)]$. We thus see that choice probability will be a logistic (nearly normal integral) function of the difference in expected utilities. This difference will increase with the differences in payoff probabilities as well as the absolute difference in utilities of the gains versus losses.

Using the method of least squares, we estimated a value of $b = 4.58$ to best fit the asymptotic choice proportions for the 12 conditions displayed earlier in Table 2.3. Inspection reveals that the expected utility model fits these data as well as does the scanning model.

The model with Luce's choice axiom generalizes easily to choice among three alternatives. Using the same value of $b = 4.58$, we predicted the choice probabilities for the Siegel & Andrews (1961) experiment that are listed in Table 2.4. Again the fit of the expected utility model to the data is excellent.

Next we consider the data from the experiment by Bower and Lloyd that used asymmetric payoffs and yielded the results displayed in Table 2.5. That experiment involved four outcomes, winning 1 or 5¢, and losing 1 or 5¢. The data from the four conditions barely suffice to estimate the four utilities and the α parameter. However, if we assume that utilities in the small range from -5¢ to $+5$¢ are approximately linear in money (i.e., $u(x) = ax + e$), then the predictions reduce to dependence on a single parameter, viz. αa. For example, in Condition 1 where subjects win or lose 5¢ after response A_1 and 1¢ after response A_2 with $\pi_1 = 1 - \pi_2 = .65$, the average utilities at asymptote should be

$$\bar{u}_1 = .65u(5) + .35u(-5)$$
$$\bar{u}_2 = .35u(1) + .65u(-1)$$

The difference is

$$\bar{u}_1 - \bar{u}_2 = .65[u(5) - u(-1)] - .35[u(1) - u(-5)]$$
$$= .30[6a] = 1.8a. \tag{7}$$

In the last line, we have substituted the assumed utility function, $u(x) = ax + e$. Exactly the same difference in expected utility arises in Condition 3 of the Bower and Lloyd experiment, which simply switched which response yielded $\pm 5\cent$ or $\pm 1\cent$. Thus, Conditions 1 and 3 should yield identical choice behavior. By similar algebra, we derive that the expected utility difference in Condition 2 of Table 2.5 should be $3a$, and that this should be identical to the difference for Condition 4. When substituted into Luce's strength equation, the predicted choice probabilities should be $p_{1,2} = 1/(1 + e^{-\alpha a(1.8)})$, for Conditions 1 and 3, and $p_{1,2} = 1/(1 + e^{-\alpha a(3.0)})$, for Conditions 2 and 4. The least-squares estimate of αa was 0.73 and the predictions are those shown in the right-hand column of Table 2.5. Obviously, the theory fits the observed choice proportions well.

Estes' Pair Comparison Experiment. We now apply the expected value model to Estes' pair comparison experiment (1966) described earlier. Recall that subjects had received many learning trials, choosing between alternative A (3 points) versus E (1 point), alternative B (3 points) versus D (2 points), and alternative C (2 points) versus F (1 point). After reaching consistent choices of the higher-valued alternative, subjects then (a) estimated the points associated with each alternative, yielding the data in Table 2.2, and (b) chose their preferred alternative in a series of 12 pair comparisons, yielding the data in Table 2.1.

A minor complication in applying the expected utility model to this experiment is that the theory yields only a single average number $(\bar{u}_i)$ for each choice alternative, whereas Table 2.2 shows a distribution of guessed-at values associated to each alternative. The expected utility theory will interpret these percentages as arising from subjects' distributing their responses over the available categories according to some outcome generalization function centered around the average utility calculated in the theory. (Estes [1966] made essentially the same generalization interpretation for the scanning model.)

Finally, we assume here that over the small range of 1 to 3 points used in this experiment, utility is linear in the number of points received. With these assumptions, we may then interpret the mean point-values (shown in the last column of Table 2.2) as linearly related to the model's average utilities for each of the alternatives. With these average utilities in hand, we may calculate the predicted choice proportions of A_i over A_j in the pair-comparison matrix of Table 2.1, using the logistic function, that is,

$$p_{i,j} = \frac{1}{1 + e^{-b(p_i - p_j)}}.$$

In this equation, p_i is the average estimated points for alternative A_i and $b = \alpha a$. Using a least-squares estimate of $b = 2.89$, we obtained the predictions shown second in parentheses in Table 2.1. The predictions generally correspond to the data. The correlation between the predicted and observed proportions is .96 (compared to .95 for the scanning model), with an average mean deviation of .027 (vs. .030 for the scanning model). Thus, we may conclude that the expected utility model is adequately describing these results.

We should point out the close correspondence of the expected utility model to Thurstone's theory for pair comparison experiments (Thurstone, 1927; Torgerson, 1958). Each theory assigns a strength to each alternative and supposes that choice of A_i over A_j is a logistic (or normal integral) function of the difference in their strengths. The expected utility theory is more constrained (and hence more falsifiable) as in this and other applications it gives a trial-by-trial learning process that arrives asymptotically at strengths bearing some simple relation to the payoffs.

Comparative Situations

We turn now to the second general class of experiments, those in which the subject receives mainly comparative feedback on each trial regarding the difference in outcomes associated with each alternative active on the trial. The paradigmatic comparative situation is the horse race (or competitive game or election) in which the feedback is typically of the form that alternative A_i (horse, sports team, candidate, etc.) beat alternative A_j in a particular race. More generally, subjects can receive feedback about the difference in performance outcomes $O_i(n) - O_j(n)$ for the two alternatives. We will suppose that this difference feedback motivates subjects to change their strength of expectations for each alternative so as to reduce the "error" or discrepancy between the A_i to A_j difference they expected versus that which they observed on this trial. Suppose we let λ_{ij} be a constant reflecting the desired absolute strength difference between the winner and the loser of a race, and $\lambda_{ij}(n)$ be the desired difference between A_i and A_j on trial n. When A_i beats A_j, then $\lambda_{ij}(n) = \lambda_{ij}$; when A_j beats A_i, then $\lambda_{ij}(n) = -\lambda_{ij}$. The crucial assumption of the model with comparative feedback is that the strengths of the two candidates in a race change on each trial according to Equations 8a and 8b:

$$s_i(n + 1) = s_i(n) + \theta[\lambda_{ij}(n) - (s_i(n) - s_j(n))] \tag{8a}$$

$$s_j(n + 1) = s_j(n) - \theta[\lambda_{ij}(n) - (s_i(n) - s_j(n))] \tag{8b}$$

In these equations, recall that $\lambda_{ij}(n)$ is positive on trials when A_i beats A_j, and is negative $(-\lambda_{ij})$ when A_i loses to A_j in a race. Note that when A_i beats A_j, Equation 8a increases $s_i(n)$ because the comparative outcome is greater than the expected difference, and Equation 8b reduces $s_j(n)$; the reverse changes occur when A_j beats A_i. These equations move the difference $s_i - s_j$ closer to where it should be to predict that trial's outcome. For example, if $s_i(n) = .6$, $s_j(n) = .3$, $\theta = .2$, and $\lambda_{ij} = 1$, then if A_i beats A_j on trial n, the new strengths for A_i and A_j would be .74 and .16; if A_j beats A_i, the new strengths would be .34 and .56. In the first case, the difference $s_i - s_j$ increases from .30 to .58 to reflect A_i beating A_j; in the second case, the difference $s_i - s_j$ decreases from .30 to $-.22$ to reflect A_j beating A_i.

Equations 8a and 8b have several curious properties. A first property, obvious in the numerical examples shown, is that the *sum* of the strengths of the two alternatives in a race remains constant from trial n to $n + 1$. Thus, if two candidates race repeatedly against one another without competing with others, then their summed strengths will remain constant. If we assume further that strengths begin, say, at zero, then the constant-sum property implies that $s_i(n) = -s_j(n)$ for all trials.

A second interesting property of Equations 8a and 8b is that they imply a simple equation for how the difference changes. Let $d_{ij}(n) = s_i(n) - s_j(n)$. Then we can derive the following relationship:

$$d_{ij}(n + 1) = d_{ij}(n) + 2\theta[\lambda_{ij}(n) - d_{ij}(n)]. \tag{9}$$

Note how the numerical examples above exemplify this relation. The implication of Equation 9 is that if A_i always beats A_j, then the strength difference $s_i - s_j$ would increase to equal λ_{ij}; if A_j always beat A_i, the difference would tend to $-\lambda_{ij}$. If A_i beats A_j on a proportion π_{ij} of their races (and neither is perturbed by involvement in other races), then their strengths will change so that their difference arrives asymptotically at $\lambda_{ij}(2\pi_{ij} - 1)$ whereas their sum stays at zero (assuming initial strengths of zero). This means that the asymptotic strengths of the two candidates will be $s_i = -s_j = \lambda_{ij}(\pi_{ij} - .5)$.

Neely's Experiment

We apply the comparative learning model to the results of Neely's political-candidates experiment. Neely's conditions were described earlier. Recall that a strong (S) and a weak (W) candidate were established in Phase 1 along with two candidates of equal but indeterminate strengths (M). The latter two are indeterminate because they won 50% of their contests against one another, and entered no other contests in Phase 1. Three of the four candidates were continued into Phase 2 of the experiment where a novel strong candidate (NS) won half his contests against S, a novel weak candidate (NW) won half his contests against W, and a novel indeterminate candidate won half his contests against M. Subjects then chose between the six candidates in the 15 possible pairings given in two test trials, one after half and another after all the trials in Phase 2.

In applying the model to the data, certain difficulties are evident. First, the strengths of candidates M and NM relative to S and W are unknown; we will guess that their strength is zero, midway between that of S and W. Second, the number of observations per test pair is small (24 for half, 48 for half), so the choice proportions have relatively large standard errors.

According to the preceding derivations, during Phase 1 the strength of the strong candidate, S, will converge to λ ($\pi - .5$), which is .30 assuming $\lambda = 1$ and the probability of candidate S winning over W is .80. Similarly the strength of candidate W will converge to $-.30$. Recall that the attractiveness of a candidate is presumed to equal $v_i = e^{\alpha s_i}$. Using the strengths of .30 and $-.30$ for S and W and an estimate of $\alpha = 3.42$, we obtain the predicted choice probabilities for the three pairs in Phase 1. (This estimate of α was actually made based on the observed choice proportions in both Phase 1 and Phase 2.) The observed preference for S over W was .91 (vs. .89 predicted), for S over M was .69 (vs. .74 predicted), and for M over W was .84 (vs. .74 predicted).

To obtain the candidates' strengths in Phase 2, we shortcut the derivations by noting that for any given pair in Phase 2 (such as S vs. NS), the sum of the two strengths on any trial will be constant and equal to the initial value of the sum at the beginning of Phase 2. This means that in Phase 2, the strengths of S and NS will always add up to .30; because these must eventually be equal (due to the 50:50 winnings of each against the other) by the end of Phase 2, we can derive that by the end of Phase 2 $s_S = s_{NS} = .15$. By a similar line of reasoning, we may derive that the strengths of the other four candidates by the end of Phase 2 will be $s_W = s_{NW} = -.15$, and $s_M = s_{NM} = 0$. These are expected asymptotic values, as a strict derivation would require estimation of θ as well.

Using these strengths and the same $\alpha = 3.42$ as before, we predict the choice proportions shown for Phase 2 in Table 2.6 where they are compared to the data from Neely's Experiment 1. The predictions generally capture the major trends in the data. Including the three Phase 1 choice proportions, the correlation between observed and predicted proportions is .88. Using a chi-square goodness-of-fit test, the hypothesis that the observed proportions are equal to the expected proportions cannot be rejected, χ^2 (16, $n = 24$) = 12.65, p > .05. Thus the

TABLE 2.6
Proportion of Choices of Row Over Column Candidate from Phase 2 of Experiment 1 of Neely (1982), Based on His Tables 1, 2, and 3[a]

	M	W	NS	NM	NW
S	71 (63)	79 (74)	50 (50)	50 (63)	79 (74)
M	—	79 (63)	46 (37)	48 (50)	63 (63)
W	—	—	33 (26)	54 (37)	44 (50)
NS	—	—	—	65 (63)	67 (74)
NM	—	—	—	—	37 (37)

[a] Prediction of the EU model are shown in parentheses.
Note. Reprinted by permission.

expected utility model provides a good account of the data from Neely's experiments.

Point-Spreads. The comparative situations considered so far are those in which subjects learn simply who beats whom in a race. However, it should be a simple matter to generalize the comparative Equations 8a and 8b to take into account the *amount* by which candidate i beats j. Recall that $O_i(n) - O)_j(n)$ is the amount by which A_i beats A_j on Trial n. Equations 8a and 8b would then alter the strengths s_i and s_j so that their strength difference comes to equal their difference when competing. That difference in competitive sports (basketball, football) is called the "point spread" and is the extent to which professional gamblers expect one team to beat another. In fact, fans gamble on whether the favored team will beat its opponent by more or less than the point spread. Similar practices are followed in golf and horse racing where it is called "handicapping." For example, a golfer who shoots a round of golf 12 strokes above the expected minimum for a given course is said to have a 12-stroke handicap; this system enables players with quite different skill levels to nonetheless play against each other competitively; each subtracts his or her handicap from the total score.

Laboratory experiments to test the point-spread version of Equations 8a and 8b are easily devised. As one example in sports-team format, team A would beat team C by relatively large margins, whereas team B would barely eke out its wins over team C. The result of such experiences in the model would be to place team A at a greater strength than B, so that in a test trial A should be predicted to beat B with a point-spread related to the difference between the average A-C and B-C point spreads. In such cases, subjects would be using the observed point spreads to infer the relative locations of the teams on the underlying competitive strength scale, much as Thurstone's law of comparative judgment uses the proportion of times team *i* beats team *j* as a derived index of the distance between their average strengths.

In a preliminary experiment by Heit, Price, and Bower (1991), subjects predicted the winner and the margin of winning ("point spread") in each of a series of 78 basketball games. The games were those actually played during a regular season (excluding postseason playoffs) by the six teams of the Atlantic division of the National Basketball Association. The teams were assigned fictitious labels to circumvent past associations of actual names to winning records. Each team played in 26 games, 5 or 6 against each of the other teams. In a final set of 15 test games which paired each team against every other, subjects predicted the point spread but received no feedback. Equations 8a and 8b were used to characterize subjects' gradual learning of the relative strengths of the teams. The best-fitting estimate of theta was .10; lambda was set each trial to the actual winning margin, and that varied from 1 to 34 in different games.

Despite the small number of trials, subjects rapidly learned the relative strengths of the teams, so that by the last 20 trials their average margin correctly predicted the winner in 75% of the games. Moreover, the model predicted subjects' average point spreads reasonably well, with a correlation of .90 between the model's and the subjects' average point spreads over the last half of training. Interestingly, the model was somewhat more accurate than were subjects in predicting the actual point spreads. In addition to fitting the training results, the model also predicted subjects' choices during the 15 test trials given after training: the model agreed perfectly with subjects on which team would be the winner; there was also a correlation of .93 between the model's and subjects' predicted point spreads over these 15 games. This is striking confirmation of the model given that it is predicting responses for single games. This preliminary experiment gave sufficiently regular results that we have been encouraged to test the model by conducting further experiments patterned after those of Neeley (1982) described previously.

Incompleteness of Our Theory

It is appropriate to note several incompletenesses or embarrassments of the theory we have proposed in the preceding pages. In this manner we point to aspects requiring more thought and theoretical efforts.

A first incompleteness is that we have not sharply characterized the class of absolute versus comparative situations. We have the two prototypes—the two-armed bandit with feedback only for the alternative chosen (absolute) versus the horse race where one only hears of the winner (comparative). But the "absolute" and "comparative" categories are in fact fuzzy and have many intermediate variants that blend into one another. The experimenter can give complete information in the two-armed bandit experiments regarding what payoff subjects would have received had they chosen the other option. We would then want to apply Equation 3 to both alternatives on such full-information trials. But isn't this example then also a case of "comparative" feedback? Similarly, in the horse race situation, one can tell subjects exactly how fast each horse ran in the race, or by how much one horse beat the other. Doesn't this then provide absolute information about each alternative's performance value? Obviously we need a sharper definition or set of criteria for deciding when to apply Equation 3 versus 8 to a given choice situation.

Beyond the issue of vagueness in our absolute/comparative distinction, the expected utility choice theory itself has had a rocky history of disconfirmations in the behavioral sciences. Some of the difficulties with the standard deterministic theory are avoided in our probabilistic choice axiom; thus, occasional inconsistencies and violations of transitivity of choice, and the like, need not upset our probabilistic version. Our theory does imply strong stochastic transitivity, that is, if $p_{12} \geq .5$ and $p_{23} \geq .5$, then p_{13} will be at least as great as the larger of p_{12} and p_{23}, and choice proportions usually exhibit this property. Our choice rule also implies the *product rule*, (see Estes, 1962); that is, for any three choice alternatives i, j, and k, the model implies that $w_{ik} = w_{ij} w_{jk}$, where $w_{ij} = p_{ij}/(1-p_{ij})$.

One problematic prediction of the comparative model in Equation 8 is that later losses against tough opponents can wipe out an advantage a team demonstrated earlier against a weak opponent. For instance, suppose in a first series of games team B beats team C 70% of the time. If team B is subsequently beaten repeatedly by one or more other teams, Equation 8 will drive down the inferred strength of team B, perhaps to a level below that of team C. On the other hand, the strength of team C established earlier will not be affected if it is not involved in the interpolated games. Thus, in a re-test of team B against C, the model could predict that C will now beat B, because the interpolated losses of B have reduced its strength. We doubt whether subjects will show similar losses upon retesting B versus C, although the issue requires empirical testing. The problem here arises due to the procedure of isolating the early B vs. C trials from the subsequent games involving B; the problem would not arise were the two sets of games to be randomly intermixed in each training cycle.

Another limitation of our comparative model is that it is only worked out for pairwise races, of two candidates at a time. But horse races, like many elections, can have many horses racing against one another, and one can record their order of finishing and even the distances between successive finishers. We would like to be able to use this ordering of inter-horse distances at the finish along with some variant of Equation 8 in order to adjust the strengths of all candidates appropriately. We are currently exploring different ways to do this. One simple solution is to treat the rank order on each race as an array of ordered pairs, so that a finishing order of B D C A in a four-horse race would be treated equivalently to the pairwise outcomes (BD), (BC), (BA), (DC), (DA), and (CA); then we would apply Equation 8 to each of these pairwise races to calculate the new strengths of the four horses. A problem with this method is that Equation 8 is noncommutative so that the final strengths would depend on the order in which the hypothetical pairwise races were considered to occur.

Concluding Comment

It is time that we bring our story to a close. We began by reviewing Estes' work on how rewards and punishments influence human behavior, specifically risky choices under uncertainty. Estes' scanning model sharply distinguished between two processes—one, the learning of which outcomes followed which responses in given situations; the other, the scanning process by which expected outcomes are implicitly generated for the choice alternatives, compared, and lead to a decision on any given choice trial. We reviewed some of the evidence Estes had marshaled in support of the scanning model, and noted that it was indeed impressive. Nonetheless, we were able to find several areas for which the predictions of the scanning model were discrepant from the data. We introduced an alternative model based on expected utility theory combined with a probabilistic choice axiom. Moreover, we distinguished between situations in which following their choices subjects receive primarily absolute versus comparative feedback. This distinction is critical as the form in which the expected utility model is applied depends on which situation is being modeled. The enhanced model was then shown to account in quantitative detail not only for the original results offered in support of the scanning model, but also and especially for those discrepant data that had proved refractory to explanation by the scanning model. We do not claim that we have explained all the relevant results on risky choices or paired comparisons that one might assemble in favor of the scanning model; a thorough review of that evidence would be a large task extending beyond the space limitations of these pages. However, the results that the utility model fits strike us as important ones to be explained.

It is with mixed feelings that we criticize and attempt to replace the scanning model. It is an elegantly simple but powerful model that makes strong, parameter-free predictions in diverse circumstances. Such models are rightfully admired and sought after in the behavioral sciences, and are modified or replaced only after the discrepancies from diverse data become glaring. It is a tribute to the insight and ingenuity of Bill Estes that a model he proposed about 30 years ago is still considered a viable theory, one that is being actively discussed, and guiding research today and probably for some years to come. The collection of such stimulating theories comprises the legacy of Bill Estes, one of the most influential learning theorists of our times.

ACKNOWLEDGEMENTS

Gordon Bower is Professor of Psychology at Stanford University. He was a colleague of Estes from 1961 to 1967 at Stanford, and they often discussed the earlier material of this chapter. Bower is supported by NIMH grant MH-13950 and AFOSR grant 87-0282. Evan Heit is a 1990 PhD from Stanford where he worked with Bower; Heit was supported by an NSF Graduate Fellowship. The authors are grateful to Doug Medin, Judy Florian, and Steve Kosslyn for comments on this chapter.

REFERENCES

Atkinson, R. C. (1962). Choice behavior and monetary payoffs. In J. Criswell, H. Solomon, & P. Suppes (Eds.), *Mathematical methods in small group processes*. Stanford, CA: Stanford University Press.

Bower, G. H., & Hilgard, E. R. (1981). *Theories of learning* (5th ed.). Englewood Cliffs, NJ: Prentice-Hall.

Bower, G. H., & Lloyd, M. (1963). *Probability learning with asymmetric payoffs and losses*. Unpublished experimental results.

Estes, W. K. (1962). Theoretical treatment of differential reward in multiple-choice learning and two-person interactions. In J. Criswell, H. Solomon, & P. Suppes (Eds.), *Mathematical methods in small group processes*. Stanford, CA: Stanford University Press.

Estes, W. K. (1966). Transfer of verbal discriminations based on differential reward magnitudes. *Journal of Experimental Psychology, 72,* 276–283.

Estes, W. K. (1969). Reinforcement in human learning. In J. Tapp (Ed.), *Reinforcement and behavior*. New York: Academic Press.

Estes, W. K. (1976a). The cognitive side of probability learning. *Psychological Review, 83,* 37–64.

Estes, W. K. (1976b). Some functions of memory in probability learning and choice behavior. In G. H. Bower (Ed.), *The psychology of learning and motivation* (Vol. 10, pp. 2–46). New York: Academic Press.

Friedman, M. P., Padilla, G., & Gelfand, H. (1964). The learning of choices between bets. *Journal of Mathematical Psychology, 1,* 375–385.

Heit, E., Price, P., & Bower, G. H. (1991). *An adaptive strength-comparison model of predicting basketball winners*. Paper presented at annual meeting of the Psychonomic Society, San Francisco.

Keller, L., Cole, M., Burke, C. J., & Estes, W. K. (1965). Reward and information values of trial outcomes in paired-associate learning. *Psychological Monographs, 79,* (12, whole No. 605).

MacCorquodale, K., & Meehl, P. (1954). Edward C. Tolman. In W. K. Estes et al. (Eds.), *Modern learning theory* (pp. 177–266). New York: Appleton-Century-Crofts.

Myers, J. L., & Suydam, M. M. (1964). Gain, cost, and event probability as determiners of choice behavior. *Psychonomic Science, 1,* 39–40.

Neely, J. H. (1982). The role of expectancy in probability learning. *Journal of Experimental Psychology: Learning, Memory, and Cognition, 8,* 599–607.

Ross, B. H., & Bower, G. H. (1981). Comparisons of models of associative recall. *Memory & Cognition, 9,* 1–16.

Siegel, S. (1961). Decision making and learning under varying conditions of reinforcement. *Annals of the New York Academy of Sciences, 89,* 766–783.

Thorndike, E. L. (1931). *Human learning*. New York: Century.

Thurstone, L. L. (1927). A law of comparative judgment. *Psychological Review, 34,* 273–286.

Torgerson, W. S. (1958). *Theory and methods of scaling*. New York: Wiley.

3 A Path Taken: Aspects of Modern Measurement Theory

R. Duncan Luce
University of California, Irvine

Let me begin by welcoming William Estes to the equivalence class of emeritus Harvard professors. I can reassure him from personal experience that it does not mean a life of idleness; of course, he has assured that himself by becoming the founding editor of *Psychological Science,* the flagship journal of the American Psychological Society.

SOME PATHS OF MATHEMATICAL PSYCHOLOGY

Although Bill Estes' and my careers have run in many parallel strands, interweaving in a number of ways, we have in fact pursued quite different intellectual paths. Figure 3.1 indicates six of the main paths (plus a small illustrative sample of names for each) that have evolved in mathematical psychology during this time. A major distinction among these paths is whether one treats the topic of study as a black box whose properties are to be captured by the mathematical model or whether one attempts to model some of the structure within the black box, sometimes to the point of resembling what neural scientists tell us about neural networks.[1] Within each are further splits that to a great extent reflect different mathematical methodologies. Some scientists, like Patrick Suppes, have pursued several of these paths; others, like me, two plus a little of a third; and still others, like Bill, have been true to one.

---

[1]A similar split is found in other sciences—quantum mechanics and relativity theory are the classic physical examples from the first part of this century. A closely related pair at the two levels is electromagnetic theory and the theory electrons. In general, each approach seems to have its role.

As a consequence of our different paths, my remarks, unlike those of many other contributors to this symposium, do not bear directly on anything that Bill has done. Rather, I describe some of the progress achieved in one of the other paths, measurement theory.

Before doing that, a comment is warranted on the relation of measurement theory to stochastic modeling and, in particular, statistics. The distinction is this: Measurement focuses on structural relations among variables, which, to use Norman Anderson's (1981) term, constitutes a cognitive algebra. Statistics, by contrast, focuses on the chance or error aspects of our measurements. In an ideal science, they would be one subject, but so far that goal has eluded us. When we don a statistical cap, we explore the adequacy of very simple algebraic models— usually purely ordinal, additive, or multiplicative ones—and we do little to understand the qualitative conditions that underlie such a representation. When we don a measurement cap, we explore the representation, but with considerable discomfort over errors of measurement because we do not really know how simultaneously to model the algebra and the errors. Achieving unity of these approaches appears to be a difficult and probably a deep problem.

ORIGINS OF MODERN DEVELOPMENTS
IN MEASUREMENT

The Campbell–Stevens debate (late 1930s and early 1940s)

Psychology, and more generally the social sciences, have found it impossible to mimic directly physical measurement that, in part at least, was based on the *counting of units*. This is familiar from length and mass measurement and had been embodied axiomatically in the model now called *extensive measurement* (Campbell, 1920, 1928; Helmholtz, 1887; Hölder, 1901; for a later treatment see chapter 3 of Krantz, Luce, Suppes, & Tversky, 1971). Coupling this failure with the powerful intuition that some psychological variables should be measured in a continuous fashion, attempts at measurement were being made outside the counting framework.

The legitimacy of these efforts was challenged by some physical scientists, notably the physicist-turned-philosopher-of-science N. R. Campbell. At his instigation, the British Association for the Advancement of Science constituted a committee to look into the issue. A number of psychologists testified, but the major counterposition was ultimately formulated by the Harvard psychophysicist S. S. Stevens (1946, 1951). The final report of the commission stated both positions (Ferguson et al., 1940). To a man, the physicists held that measurement means just one thing: It must be possible to combine entities exhibiting the attribute to be measured into an entity that also exhibits the same attribute, and

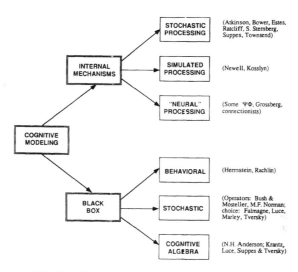

FIG. 3.1. Some paths in cognitive model building.

this combining operation is used to count the number of units approximately equal to the entity to be measured. Exactly how averaging representations, which certainly play a role in physics, were to be encompassed in this formulation was not made clear. In contrast, Stevens (and other psychologists) took the position that no one method is necessary in order for something to be classed as measurement, but rather it is a question of the degree of uniqueness achieved in the numerical representation of some body of empirical information. The key was the *scale type* as embodied in the degree of scale invariance, such as up to the specification of a unit, as in length and mass. Stevens (1946, 1951) cited five types: nominal, ordinal, interval, ratio, and absolute. The former differs from all of the others in that no ordering is involved, and today it is no longer treated as a significant part of the measurement taxonomy. Stevens (1959) later added another scale type, log-interval, but from the modern point of view that type is not really distinct from interval scales.

So far as I can tell, Stevens' taxonomy grew out of an examination of what had arisen in physics.[2] For example, on p. 34 of Stevens (1959) we find "The five scales listed . . . would seem to exhaust the possibilities, at least those of scientific interest, except perhaps for the class of scales on which no transformations would be possible. These would be ratio scales having in some sense or

---

[2]This examination first took place during an interdisciplinary faculty seminar at Harvard during the late 1930s. The mathematician George Birkhoff appears to have been highly influential in bringing out the significance of groups of transformations in this classification.

other a natural unit. . ." The Moh scale of hardness was a practical example of ordinal measurement; temperature (before the discovery of absolute zero) and time were examples of interval measurement; length, mass, duration (= time interval), and charge, of ratio measurement; and probability, perhaps, of absolute measurement.

Developing Evidence that Campbell Was Too Narrow (1950–1960s)

Although physics offered numerous examples of measurement that did not fall under Campbell's narrow filter, and he was obliged to devise his concept of derived measurement to accommodate them, it was not until somewhat later that it became very clear (at least to behavioral scientists) that his dictum was simply too narrow. Perhaps the single most telling example was the development of *additive conjoint measurement* (Debreu, 1960; Luce & Tukey, 1964) in which no explicit combining operation was evident in the primitives.[3] Rather, there were tradeoffs between two (or more) variables, each of which affect the attribute to be measured, and these tradeoffs can, under appropriate circumstances, be used to construct an indirect scheme for counting units. The situation described by this model was not only of considerable interest in the behavioral and social sciences, where such tradeoffs had been widely studied in learning, motivation, and perception, but it was also the actual basis of much physical measurement. Indeed, a careful examination of the entire structure of physical quantities showed that it consisted of a mix of conjointly and extensively measured variables that are interconnected by means of certain qualitative distribution laws, with neither seeming to have primacy over the other (see Krantz et al., 1971, ch. 10; and Luce, Krantz, Suppes, & Tversky, 1990, ch. 22).

In terms of Stevens' scale types, such conjoint measurement on continuous domains resulted in interval measurement when the additive representation (using all of the real numbers) is employed or log-interval when the multiplicative representation (on positive reals) is employed. The latter arises from the former by an exponential transformation.

Typically, physicists use multiplicative representations on the positive reals, as in

$$KE = \tfrac{1}{2}mv^2,$$

but there are exceptions such as dB measures and entropy. Economists and psychologists more often use an additive representation on all of the reals; for

---

[3]The general ideas formalized in conjoint measurement were explicitly discussed by Archimedes, but they were largely ignored during the formal development of modern measurement theory. In particular, no mention of them is found in Campbell's books.

example, utility models and many ANOVA models, but again there are exceptions such as the current attention in statistics to log-linear models.

I mention several other developments more briefly. Closely related to conjoint measurement was that of *difference measurement* (Debreu, 1958; Suppes & Winet, 1955) in which the qualitative data are judgments about the ordering of stimulus pairs according to some attribute, such as similarity. Again, in continuous domains this results in interval scales, and it serves as a building block for some models leading to geometric representations.

A different direction continues to involve an operation of combining, but instead of being positive in the sense that x combined with y is larger than both x and y, as is true of extensive measurement, the combined value lies between them. This is known as *intern* (or *intensive*) *measurement* (Aczél, 1948; Pfanzagl, 1959; von Neuman & Morgenstern, 1947). It leads to a representation in terms of some form of averaging, and in sufficiently rich contexts it usually results in interval scales but sometimes in ratio scales (see the last section).

So far, the examples have all fallen within Stevens' taxonomy, but there were even exceptions to that. Two of the most notable were Coombs' (1950) *unfolding* model for preferences and my (Luce, 1956) attempt to weaken the order axiom so that the strict order remains transitive but indifference need not be. These *semiorders* captured the qualitative notion of a threshold and under certain circumstances could be represented that way, but the scale type was never fully characterized. For recent treatments, see Fishburn (1985) and Suppes, Krantz, Luce, & Tversky, (1989).

Perplexities

So, by the mid1960s we found ourselves with a number of examples, most, but by no means all of which fell within Stevens' scheme. Moreover, despite the fact that direct counting of units was not involved, all of the examples involved addition one way or another in the representation. We clearly did not understand very well the full extent of Stevens' claim. For example, at that time the following questions were unresolved, and most psychologists suspected they would not soon be solved.

- What scale types are there other than Stevens' ordinal, interval, ratio, and absolute?
- Among all of them, whatever they are, why do Stevens' four types seem so pervasive?
- Just how crucial is addition in measurement representations?
- In particular, are nonadditive representations of any theoretical or scientific significance?

MODERN DEVELOPMENTS IN MEASUREMENT

Nonadditive Structures (1976–1985)

Mathematically, it is easy to envisage potential nonadditive conjoint representations. Consider, for example,

$$F(x,y) = x + y + \begin{cases} Cxy, & \text{if } x \geq 0, y \geq 0 \\ 0, & \text{if } xy < 0 \\ Kxy, & \text{if } x \leq 0, y \leq 0 \end{cases} \tag{1}$$

and order stimuli by:

$$(a,x) \succeq (b,y) \text{ if and only if } F(a,x) \geq F(b,y).$$

The question raised by Narens and Luce (1976) was: What properties need an ordering satisfy in order for it have some sort of nonadditive representation, for example, the one just described? It turns out that this question concerning conjoint structures, where it arises very naturally, can be reduced, when the domain of the conjoint structure is suitably rich, to a comparable question about an ordering and an operation of combining. And we gave suitable axioms on an operation to have a non-additive numerical representation. Basically all we did was drop associativity[4] from the axioms of extensive measurement.

But in that paper we failed to characterize the scale type. We were able to show that, like extensive measurement, if one value of the representation is prescribed, then the representation is uniquely determined. But we were not able to state the family of transformations that took one representation into another on the same numerical structure. This was somewhat perplexing, and only later did the reasons for our difficulty become apparent.

The breakthrough came when Michael Cohen, in his second-year graduate research paper at Harvard, proved that the scale type was either ratio or a proper subgroup of the similarity transformations of the ratio case. That striking result is included in Cohen & Narens (1979), where they also showed that the ratio case has a particularly simple representation. Later Luce & Narens (1985) extended this result to include additional scale types, including the interval scale case of intensive measurement. The basic form of the representation is that there exists a function f from the positive reals onto itself such that f(x) is strictly increasing and f(x)/x is strictly decreasing in x, and the numerical operation $\otimes$ is given by:

$$x \otimes y = yf(x/y).$$

---

[4]If $\bigcirc$ denotes the operation, associativity says that groupings of the binary operation over three or more variables does not matter, that is, for all elements, x, y, and z, $x \bigcirc (y \bigcirc z)$ is equivalent to $(x \bigcirc y) \bigcirc z$.

Note that this form is invariant under ratio scale transformations because:

$$rx \otimes ry = ryf(rx/ry) = r(x \otimes y).$$

Detailed summaries of these developments can be found in chapter 19 of Luce *et al.* (1990) and Narens (1985).

Subconclusions

The examples that were developed during this period made clear that formal theories of measurement—formal in the sense of axiomatizing an ordering over a structure and finding a numerical representation of that information—need not end up with addition or multiplication playing any role whatsoever in the representation. Yet, addition and multiplication continued to play an explicit role in two of the stronger Stevens scale types, namely, interval and ratio. In the interval case the admissible transformations of a representation ϕ are of the form:

$$\phi' = r\phi + s,$$

and obviously both addition and multiplication are involved.

So the question arose as to the degree to which this is a deep aspect of measurement and the degree to which it is superficial.

GENERAL THEORY OF SCALE TYPES (1980s)

Symmetries

Suppose a structure $\mathcal{S}$ has a ratio scale representation. This means that if ϕ is a numerical representation (isomorphism) of the structure $\mathcal{S}$ with a numerical structure $\mathcal{N}$ then $r\phi$ is an equally good representation in the same structure. This is diagrammed in Fig. 3.2. Note that the mapping $\alpha = \phi^{-1}r\phi$ is an isomorphism between the structure $\mathcal{S}$ and itself. Physicists refer to such an isomorphism as a *symmetry* of the structure; mathematicians speak of an *automorphism* (= self-isomorphism).

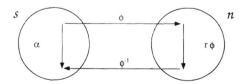

FIG. 3.2.   The relations among numerical isomorphisms, admissible transformations, and symmetries for the ratio case.

Properties of Symmetries

The new focus that arose in the 1980s, initiated by Narens (1981a, b), is on the symmetries of a structure rather than on the structure itself and its representations. In a sense, the latter comes for free provided that the symmetries are sufficiently rich.

Narens posed the problem of classifying the symmetries and he introduced the following two key ideas. A structure is *homogeneous* if for each pair of points in the structure, some symmetry maps the one point into the other.[5] Ratio, interval, and ordinal scale structures whose ranges are onto an infinite interval of the real numbers are all homogeneous. Homogeneity excludes structures having an element that is structurally distinct from the others. Examples are structures with an identity (or zero) element, a maximum element, or a minimum element.

The other concept is that of a structure being *finitely unique* in the sense that for some integer N, whenever two symmetries agree at N points then they are, in fact, the same transformation. Another way of looking at the concept is that whenever the values of N distinct points are specified, the representation is uniquely determined. Both ratio and interval structures are finitely unique (with $N=1$ and 2, respectively). Ordinal structures are not finitely unique because it takes countably many values to specify a strictly increasing function completely.

Types of Scale Types on the Continuum

Consider structures whose ordering forms a continuum.[6] Then in terms of the concepts of homogeneity and finite uniqueness we may distinguish three quite distinct possibilities:

1. The structure is not finitely unique. This category includes ordinal cases as well as others for which the classification is not yet complete.
2. The structure is both homogeneous and finitely unique. Three distinct cases exist: ratio, interval, and ones lying between these two; e.g., $\phi' = k^n\phi + t$, where $k > 0$ is fixed, n ranges over all integers, and t over all reals.
3. The structure is not homogeneous. Here the classification, although formally complete, is not yet well understood except for some important special cases (see following).

---

[5]In group theory this is usually called "transitivity," but in the measurement context that usage is all-too-likely to generate confusion. The term "homogeneity" is reasonable in the sense that it means structurally that points are indistinguishable, which is pretty much the everyday meaning of the term. It is also the term used in a similar way in geometry.

[6]This means that the ordered domain is order isomorphic to the real numbers together with their natural order.

Thus, among the class of homogeneous and finitely unique structures nothing is possible other than ratio, interval, and cases in which the admissible transformation are a proper subset of the affine transformations of the interval case and properly include all of the similarity transformations of the ratio case. So far, no concrete example of this in-between scale type seems to have arisen in measurement. The reasons underlying Stevens' classification are now clear.

As I mentioned, Narens initiated the problem and achieved a partial solution; T. M. Alper, then a mathematics undergraduate at Harvard, solved it completely as his thesis, which was published as Alper (1987).[7]

Continued Role of Additivity

A symmetry is called a *translation* if either it is the identity, in which case every point is fixed in the sense of mapping into itself, or no point whatsoever is fixed. Among the affine transformations of interval scale measurement, the translations are of the form $\tau(x) = x + t$.

The key to the Alper-Narens result is in showing that:

(i) The set of all translations are homogeneous in the sense that given any two points, some translation maps the one point into the other.

(ii) The translations look like real numbers under addition, for example, $\sigma(x) = x + s$ and $\tau(x) = x + t$ imply $\tau(\sigma(x)) = x + s + t$.

The proof works equally well if (ii) is replaced by:

(ii') The translations are isomorphic to a subgroup of the additive numbers.

Indeed, as Luce (1986, 1987) showed:

- In the homogeneous case, (i) and (ii') provide a general recipe for measurement in the sense that given an axiom system one need only establish that (i) and (ii') are true.

- When that is true, then it is easy to construct an explicit numerical representation in which the translations appear as multiplication by a positive constant. These representations are referred to as *unit representations*.

- These structures are exactly those that can be appended to the existing measures of physics without disrupting the nice patterns that are exploited in dimensional analysis, namely, the relations among the representations as products of powers, as is evidenced in the pattern of physical units such as gram-centimeters/second^2.

---

[7]He received the 1989 Young Investigator Award of the Society for Mathematical Psychology for this paper.

Some Directions Currently Being Followed

1. The small number of homogeneous, finitely unique scale types invites an attempt to classify fully the structures of each type. Detailed results are known for binary conjoint structures, that is, orderings of two factors, and for structures with a binary operation (Luce & Narens, 1985). Some of this is illustrated later in the applications.

2. Results exist about nonhomogeneous structures with an operation and with singular points (zero, maxima, or minima) but that are homogeneous between adjacent singular points (Luce, 1991b). Examples of nonhomogeneous structures are:

   • Relativistic velocity, for which both a maximum (the velocity of light) and a minimum (which is also a zero, namely no relative motion) exist.
   • Prospect theory (Kahneman & Tversky, 1979; Tversky & Kahnneman, 1990), which is a generalization of expected utility theory in which there is a distinctive zero, the status quo, that divides consequences into gains and losses.
   • Continuous probability, with both a maximum (the universal event) and a minimum (the null event), fails to be homogeneous between them.

3. A comprehensive theory of meaningful propositions within a measurement structure has been under development for some years. Stevens (1951) argued that invariance of statistical propositions under admissible scale transformations was required for these propositions to be meaningful.[8] Luce (1978) showed that the physically important principle of dimensional invariance, which underlies the method of dimensional analysis that is used in physics, engineering, and to a lesser degree in biology, is merely a special case of demanding that physical laws be invariant under the symmetries of the structure of physical quantities. More recently, Narens (in preparation) has been exploring the concept from a deep logical perspective, showing precisely the interplay of meaningfulness and invariance ideas that have arisen in geometry, physics, mathematics, and measurement.

APPLICATIONS

Philosophically the results I have described are satisfying—one feels that a previously unexplored portion of the world (or, more apt today, the solar system) has been partially mapped. But aside from increased clarity, are the results good

---

[8]Unfortunately, he cast his point mainly in terms of statistics per se rather than the propositions in which they occur. Doing so fueled a confused controversy that probably is still not settled for everyone. See Luce et al. (1990), Section 22.6.

for anything? I think so, and to illustrate in what way I describe briefly two areas of application, namely, psychophysical matching and the theory of choices among uncertain alternatives.

Psychophysical Matching

Consider the class of procedures where a person "matches" stimuli from one physical modality to those from another—light to sound, length to weight, and so on—but does so under instructions to maintain stimulus ratios. The most obvious example is the cross-modal matching of Stevens (1957), but there are others that are less obvious. For example, one can "match" within a single domain as follows: Define M to be the match between intensity I and the intensity increment $\Delta(I)$ that makes I and $I + \Delta(I)$ "just noticeably different."

Luce (1959) stated a theory of matching that postulated that an admissible change in the representation of the stimulus being matched should reappear as an admissible change in the representation of the matching stimulus. Indeed my claim was that this can be considered a principle of theory construction akin to dimensional invariance, but this contention was challenged by Roozeboom (1962a, b; Luce, 1962). One issue is: Why should the scientists' method of representing physical information about the stimuli be of any relevance to the person doing the matching? The major conclusion of that theory was that such matching should be represented by power functions of the stimulus representations. This was of interest because it accorded (approximately) with the empirical results of magnitude estimation and cross-modal matching.

An alternative, more substantive theory was proposed by Krantz (1972) that was based, in part, on then unpublished ideas of Shepard (published later in 1978 and 1981). In his "relation theory" it is assumed that matches are always ratio judgments, implicit if not explicit. This also led to power function interconnections which, however, were somewhat obscured in this theory by an undetermined monotonic transformation. So relation theory seems to me both too strong in its supposition that there are implicit standards and, possibly, too weak in the conclusions reached.

Another approach, one that escapes the weaknesses of each of the earlier attempts, is based on exploring the hypothesis that the psychology embodied in a matching relation M is compatible with the physics embodied in the domains being matched. This can be formulated precisely in terms of the translations of the physical structures. A matching relation M is said to be *translation consistent* if associated with each translation τ of the independent variable, there is a translation σ_τ of the dependent variable that maintains the match, that is,

$$xMu \Leftrightarrow \tau(x)M\sigma_\tau(u).$$

Suppose the structures being matched have unit representations φ and φ. Then Luce (1990) showed that for some positive constants k and β,

$$xMu \Leftrightarrow \varphi(u) = k\varphi(x)^\beta.$$

For the jnd example, it says that $\Delta(I) = kI^\beta$, which includes Weber's law ($\beta = 1$) as a special case as well as the "near-miss to Weber's law" in pure tones (Jesteadt, Weir, & Green, 1977; McGill & Goldberg, 1968).

Other results are given concerning the class of relations having the same exponent and the relation of matching to ratio judgments.

In summary, then, this theory says that if the psychology is nicely compatible with the physics, then matching corresponds to a power function relation in terms of the unit structure representations. The compatibility is not forced, however, and it is an empirical matter, not one of units of representations, to decide if it obtains.

Utility Theory

Choices among uncertain alternatives provide a rich structure to which these measurement ideas can also be applied. The major primitives are consequences, events, finite gambles, and preference orderings. In particular, let C denote a (dense) set of *pure* consequences, such as sums of money received or lost. Let $\mathcal{E}$ denote a (dense) set of *chance events. A finite gamble* is simply an assignment of a pure consequence to each member of a finite partition of an event. For example, suppose the underlying event is the throw of a die, and the partition is into the subevents $\{1,2\}$, $\{3,4,5\}$, and $\{6\}$, then assigning the monetary consequences $-\$5$, $\$2$, and $\$10$, respectively, to these subevents is a finite gamble. Further, let the *domain of gambles, $\mathcal{G}$,* consist of all finite gambles and all those that can be formed inductively by substituting finite gambles for pure consequences, and let $\gtrsim$ denote a *preference ordering* over $\mathcal{G}$.

Classical Subjective Expected Utility (SEU) Theory

Since von Neuman and Morgenstern (1947) and Savage (1954), the major normative theory is *subjective expected utility* or, more briefly, SEU, which is based on four major tenants of rational behavior for choosing among uncertain alternatives (see Fishburn, 1982):

1. Preferences are *transitive*— $f \gtrsim g$ and $g \gtrsim h$ imply $f \gtrsim h$ —and *connected* — either $f \gtrsim g$ or $g \gtrsim f$.
2. Preferences are *monotonic:* If g' is generated from g by replacing a consequence of g by a more preferred consequence, then $g' \gtrsim g$.
3. *Independence of a common consequence:* Suppose g and h are binary gambles with consequences $x > y$. From g' from g and h' from h by changing y to x over a common subevent. Then $g \gtrsim h$ iff $g' \gtrsim h'$.
4. *Universal accounting equivalences:* For any two gambles, no matter how

structured into successive subgambles, if they reduce to the same normal form, that is, yield the same outcomes under the same chance events (ignoring the order of occurrence of sequences of independent events), then they are judged indifferent in preference.

With suitable assumptions about the richness of the domain, these properties are equivalent to the existence of a preference order-preserving function U over $\mathcal{G}$ and a probability measure P over $\mathcal{E}$ such that U(g) equals the expectation of U calculated for the several consequences of g's relative to the probability measure P. (The term "subjective" of SEU reflects the fact that the expectation is calculated in terms of a decision-induced probability P, rather than an objective one, as was assumed in the original von Neumann and Morgenstern presentation.) Moreover, each of the rationality axioms is implied by the SEU representation.

Troubles With SEU

There are real and pseudo problems with SEU. One of the pseudo problems that tends to afflict cognitive psychologists is the claim that the average person surely cannot mentally calculate expectations, and so a priori the theory must be wrong. This has about the same merit as saying that the theory of physical dynamics must be wrong because a chunk of matter does not embody enough computing power to solve the differential equation describing its motion. Such assertions confuse the scientist's description of the behavior with the source of that behavior.

Turning to the real empirical issues, I mention four.

Universal accounting seems unlikely to be descriptive; however, there are surprisingly few data devoted to showing that it is wrong. Models that attempt to limit its scope can be viewed as an aspect of bounded rationality, a term introduced by Simon (1955, 1978).

Some experiments raise doubts about transitivity and monotonicity. In my opinion their significance has probably been either exaggerated and/or is in some measure a misinterpretation of the data. For more detailed discussion, see Luce (1992). Certainly both properties are normatively compelling to many scientists and decision makers.

Common consequence is highly suspect for uncertain events—those for which there is no known objective probability—as was first pointed out by Ellsberg (1961). Despite its apparent reasonableness, people who violate it do not generally alter their choices to conform to it when the violation is pointed out. This contrasts sharply with their desire to resolve violations of monotonicity and, especially, transitivity.

SEU makes no distinction between gains and losses, which is universally recognized to be behaviorally significant for decision makers. Attempts have been made to deal with this in terms of the shape of the utility function, but since in SEU utility is an interval scale measure, such discussions do not really make any sense.

So, I turn to recent attempts to modify the theory, of which there are two major types.

Rank-Dependent (RD) Utility

The rank-dependent theories, which originated with Quiggin (1982) and by now come in a number of variants (see Luce, 1988 for a listing), entail a representation that, like SEU, involves a weighted average of the U-values of each consequence where, however, the weights depend both upon the event giving rise to the consequence, as in SEU, and the rank position of that consequence relative to the other consequences in the gamble.

As an example of such a representation, consider the binary gamble $(x,E;y, \neg E)$ where $\neg E$ means "not E." Then the *rank-dependent* (RD) representation takes the form:

$$U(x,E;y,\neg E) = \begin{cases} U(x)S_>(E) + U(y)[1 - S_>(E)], & \text{if } x > y, \\ U(x), & \text{if } x \sim y, \\ U(x)S_<(E) + U(y)[1 - S_<(E)], & \text{if } x < y. \end{cases}$$

where the weights, S, depend not only on the event E but also on the preference order between the two consequences.

The rank-dependent representation implies monotonicity and transitivity of preference; it need not imply independence of common consequences because the weights, unlike probabilities, need not be additive over disjoint events; and most importantly it does not imply most of the accounting equivalences. The key one that it does imply is called *event commutativity:* Suppose a gamble is conducted in two stages, each of which rests on running an independent chance experiment. Suppose, further, that x is the consequence when events E and F both occur, one in the first experiment and the other in the second independent experiment, and y is the consequence otherwise. The assertion is that the decision maker is indifferent as to the order in which these events are carried out. Because event commutativity is a major implication of rank dependent theories, Alan Brothers (1990), a recent PhD of the University of California-Irvine, has studied it empirically. It appears to be satisfied by, perhaps, a quarter of student subjects, but is systematically violated by others who seem to be using simplifying heuristic rules.

In the context of binary gambles, Luce and Narens (1985) proved that rank dependence is the only possible generalization of SEU that retains transitivity and monotonicity of preference, event commutativity, and the existence of an interval scale representation.

Although rank dependence accommodates many of the empirical anomalies of SEU, it fails to take into account the special status of the status quo (or more generally of some aspiration level) and of the asymmetries of gains and losses relative to that point. To deal with that one has to abandon purely homogeneous theories and to develop those that have at least one highly distinctive point.

Rank- and Sign-Dependent (RSD) Utility

So far three related efforts in this direction have appeared. Edwards (1962) commented on the possibility of ratio scale measurement of utility when the weights are not additive and discussed the importance of the status quo in defining gains and losses. Kahneman and Tversky (1979) gave a *rank- and sign-dependent* (RSD) utility representation that they called *prospect theory*. Its domain of application was restricted to gambles with at most one positive, one negative, and one null consequence. Tversky and Kahneman (1990) generalized it to finite gambles. Recently I have been working on a different axiomatic, measurement theoretic generalization of prospect theory leading to a similar representation (Luce, 1991a; Luce & Fishburn, 1991). What sharply distinguishes our theories from the earlier ones (an exception is Pfanzagl [1959]) is not only an explicit focus on gains and losses, but the introduction of a binary operation $\oplus$ of joint receipt: If f and g are gambles, f$\oplus$g denotes the receipt of both f and g. Having this allows one to formalize some of the editing processes discussed informally by Kahneman and Tversky.

In the simplest version of the theory, this operation is assumed to be extensive in character and so there is a utility function U that is additive[9] over $\oplus$, i.e.,

$$U(x \oplus y) = U(x) + U(y).$$

To deal with the asymmetries of gains and losses, note that any gamble g is formally equivalent to a composite gamble consisting of three subgambles:

a gamble, g^+, of only gains, which is conditional on a subevent E(+) occurring,

a gamble of no change, e, conditional on a subevent E(0) occurring,

a gamble, g^-, of only losses, which is conditional on a subevent E(-) occurring,

where E(+)$\cup$E(0)$\cup$E(-) = E is the event on which g is conditional.

A major axiom of the theory is the (rational) accounting assertion that people perceive a gamble as indifferent to its formally equivalent composite gamble of gains and losses, that is,

$$g \sim (g^+, E(+); e, E(0); g^-, E(-)).$$

---

[9]A somewhat weaker assumption, like Equation 1, is used in Luce and Fishburn (1991), but it is simpler here to exposit just the additive case.

A second major axiom of the theory is that this composite gamble can be recast as the joint receipt of the gamble of gains pitted against no change and the gamble of losses pitted against no change. The assumption is that decision makers (irrationally) perceive the gamble g as equivalent to this joint receipt, that is,

$$g \sim (g^+, E(+); e, E(0) \cup E(-)) \oplus (g^-, E(-); e, E(0) \cup E(+))$$

And a third axiom is an editing one that strikes me as being rational in character, although it is not usually listed among the axioms of rationality because $\oplus$ is not usually among the primitives.[10] The axiom is: A finite gamble that is composed entirely of gains (losses) is perceived as indifferent to the joint receipt of the smallest gain (loss) and the modified gamble in which each consequence is reduced by that smallest gain (loss).

RSD Representation

The assumptions just described, along with transitivity and monotonicity, lead to a representation U involving two sets of weights, W^+ and W^-, both assuming values in the unit interval $[0,1]$ and defined over all the events. This representation is both rank and sign dependent in the following sense: Let $g \sim (g^+, E(+); e, E(0); g^-, E(-))$. Then

(i) $U(g) = U(g^+)W^+[E(+)] + U(g^-)W^-[E(-)]$.

(ii) $U(g^+)$ has an ordinary RD representation with weights that also depend on $+$.

(iii) $U(g^-)$ has an ordinary RD representation with weights that also depend on $-$.

(iv) U is a ratio scale.

The ratio scale feature arises from two facts: that $U(e) = 0$ and that in property (i) the weights do not, in general, add to 1, that is,

$$W^+[E(+)] + W^-[E(-)] \neq 1.$$

This nonadditivity of the weights arises both because $U(e) = 0$ wipes out any weight assigned to E(0) and because W^+ is generally not closely related to W^-.

Kahneman and Tversky's (1979) prospect theory is the special case of RSD applied to gambles with only one gain and one loss and some added constraints

---

[10]As noted earlier, Pfanzagl (1959) is the only exception that I know of. He studied the axiom stated here, which he called the *consistency principle*, in the binary case but without distinguishing between gains and losses, as Fishburn and I do.

among the weights. Prospect theory has accounted for a good deal of anomalous data, and Tversky and Kahneman (unpublished) have tested, with apparent success, the more general representation. Among the issues to be studied further empirically, three stand out:

• In the domains of all gains and all losses, does event commutativity hold, as the theory says it must because the representation is rank dependent in these domains?

• Rank dependence arises primarily from the editing axiom, and so that axiom needs to be studied directly.

• The connection between gains and losses is achieved into two ways, both of which need careful examination. One is the additive representation of the joint receipt operation $\oplus$, although that can be weakened somewhat to the form of Equation 1. The other, and key, connection is the nonrational decomposition of a gamble of gains and losses into the joint receipt of two gambles, the one of gains pitted against the null consequence and the other of losses pitted against the null consequence. One study, Slovic and Lichtenstein (1968), supports the latter assumption. But both assumptions require considerably more empirical study and possible modification.

I consider the work done to date to be just a beginning of research on ratio scale theories of utility. We have explored only those ratio theories that involve a heavy dose of averaging, but as I have shown for general n-ary operations (Luce, unpublished) a large family of nonadditive, nonaveraging possibilities exists. Little is known about axiomatizing specific members of that family, but the need to restrict possibilities certainly invites the formulation of new behavioral axioms. Additional general theory about nonhomogeneous outcomes, especially those that are homogeneous between singular outcomes, is needed as input to these more psychological applications.

ON PATHS NOT TAKEN

In reflecting on the different paths taken by Bill and me and others, I was reminded of the closing lines of Robert Frost's poem *The Road Not Taken*. The "I" in the poem could be either of us as, in the 1950s, I'm not sure which path would then have been deemed the one less traveled.

> Two roads diverged in a wood, and I—
> I took the one less traveled by,
> And that has made all the difference.

ACKNOWLEDGMENTS

Preparation of this chapter was supported in part by National Science Foundation Grant IRI-8996149 to the University of California, Irvine. Louis Narens, who has been a major figure in the developments reported here, provided useful comments on this paper as did Patrick Suppes, who served as a referee.

REFERENCES

Aczél, J. (1948). On mean values. *Bulletin of the American Mathematical Society, 54,* 392–400.

Alper, T. M. (1987). A classification of all order-preserving homeomorphism groups of the reals that satisfy finite uniqueness. *Journal of Mathematical Psychology, 31,* 135–154.

Anderson, N. H. (1981). *Foundations of information integration theory.* New York: Academic Press.

Brothers, A. J. (1990). *An empirical investigation of some properties relevant to generalized expected utility.* Doctoral dissertation, University of California, Irvine, CA.

Campbell, N. R. (1920). *Physics: The elements.* Cambridge: Cambridge University Press. (Reprinted as *Foundations of science: The philosophy of theory and experiment.* New York: Dover, 1957.)

Campbell, N. R. (1928). *An account of the principles of measurement and calculation.* London: Longmans, Green.

Cohen, M., & Narens, L. (1979). Fundamental unit structures: A theory of ratio scalability. *Journal of Mathematical Psychology, 20,* 193–232.

Coombs, C. H. (1950). Psychological scaling without a unit of measurement. *Psychological Review, 57,* 145–158.

Debreu, G. (1958). Stochastic choice and cardinal utility. *Econometrica, 26,* 440–444.

Debreu, G. (1960). Topological methods in cardinal utility theory. In K. J. Arrow, S. Karlin, & P. Suppes (Eds.), *Mathematical methods in the social sciences, 1959* pp. 16–26, Stanford, CA: Stanford University Press.

Edwards, W. (1962). Subjective probabilities inferred from decisions. *Psychological Review, 69,* 109–135.

Ellsberg, D. (1961). Risk, ambiguity, and the Savage axioms. *Quarterly Journal of Economics, 75,* 643–669.

Ferguson, A. (Chairman), Meyers, C. S. (Vice-chairman), Bartlett, R. J. (Secretary), Banister, H., Bartlett, F. C., Brown, W., Campbell, N. R., Craik, K. J. W., Drever, J., Guild, J., Houstoun, R. A., Irwin, J. O., Kaye, G. W. C., Philpott, S. J. F., Richardson, L. F., Shaxby, J. H., Smith, T., Thouless, R. H., & Tucker, W. S. (1940). Quantitative estimates of sensory events. *The Advancement of Science. The Report of the British Association for the Advancement of Science. 2,* 331–349.

Fishburn, P. C. (1982). *The Foundations of Expected Utility.* Dordrecht: Reidel.

Fishburn, P. C. (1985). *Interval Orders and Interval Graphs.* New York: Wiley.

Helmholtz, H. von (1887). Zählen und Messen erkenntnis-thoretisch betrachet. *Philosophische Aufsätse Eduard Zeller gewidmet,* Leipzig. English translation by C. L. Bryan, (1930). *Counting and measuring.* Princeton, NJ: Princeton University Press.

Hölder, O. (1901). Die Axiome der Quantität und die Lehre vom Mass. Sächsische Akademie Wissenschaften zu Leipzig, Mathematisch-Physische Klasse, 53, 1–64.

Jesteadt, W., Wier, C. C., & Green, D. M. (1977). Intensity discrimination as a function of frequency and sensation level. *Journal of the Acoustical Society of America, 61,* 169–177.

Kahneman, D., & Tversky, A. (1979). Prospect theory: An analysis of decision under risk. *Econometrica, 47*, 263–291.

Krantz, D. H. (1972). A theory of magnitude estimation and cross-modality matching. *Journal of Mathematical Psychology, 9*, 168–199.

Krantz, D. H., Luce, R. D,. Suppes, P., & Tversky, A. (1971). *Foundations of measurement* (Vol. I). New York: Academic Press.

Luce, R. D. (1956). Semiorders and a theory of utility discrimination. *Econometrica, 24*, 178–191.

Luce, R. D. (1959). On the possible psychophysical laws. *Psychological Review, 66*, 81–95.

Luce, R. D. (1962). Comments on Rozeboom's criticisms of 'On the possible psychophysical laws.' *Psychological Review, 69*, 548–551.

Luce, R. D. (1978). Dimensionally invariant numerical laws correspond to meaningful qualitative relations. *Philosophy of Science, 45*, 1–16.

Luce, R. D. (1986). Uniqueness and homogeneity of ordered relational structures. *Journal of Mathematical Psychology, 30*, 391–415.

Luce, R. D. (1987). Measurement structures with Archimedean ordered translation groups. *Order, 4*, 165-189.

Luce, R. D. (1988). Rank-dependent, subjective expected-utility representations. *Journal of Risk and Uncertainty, 1*, 305–332.

Luce, R. D. (1990). "On the possible psychophysical laws" revisited: Remarks on cross-modal matching. *Psychological Review, 97*, 66–77.

Luce, R. D. (1991). Rank- and sign-dependent linear utility models for binary gambles. *Journal of Economic Theory, 53*, 75–100.

Luce, R. D. (1992). Where does subjective expected utility fail descriptively? *Journal of Risk and Uncertainty, 5*, 5–27.

Luce, R. D. (unpublished). Generalized concatenation structures that are homogeneous between singular points. Manuscript submitted for publication.

Luce, R. D., & Fishburn, P. C. (1991). Rank- and sign-dependent linear utility models for finite first-order gambles, *Journal of Risk and Uncertainty, 4*, 29–59.

Luce, R. D., Krantz, D. H., Suppes, P., & Tversky, A. (1990). *Foundations of Measurement* (Vol. III). New York: Academic Press.

Luce, R. D., & Narens, L. (1985). Classification of concatenation measurement structures according to scale type. *Journal of Mathematical Psychology, 29*, 1–72.

Luce, R. D,. & Tukey, J. W. (1964). Simultaneous conjoint measurement: A new type of fundamental measurement. *Journal of Mathematical Psychology, 1*, 1–27.

McGill, W. J., & Goldberg, J. P. (1968). A study of the near-miss involving Weber's law and pure-tone intensity discrimination. *Perception & Psychophysics, 4*, 105–109.

Narens, L. (1981a). A general theory of ratio scalability with remarks about the measurement-theoretic concept of meaningfulness. *Theory and Decision, 13*, 1–70.

Narens, L. (1981b). On the scales of measurement. *Journal of Mathematical Psychology, 24*, 249–275.

Narens, L. (1985). *Abstract measurement theory*. Cambridge, MA: MIT Press.

Narens, L. (in preparation). *A theory of meaningfulness*.

Narens, L., & Luce, R. D. (1976). The algebra of measurement. *Journal of Pure and Applied Algebra, 8*, 197–233.

Pfanzagl, J. (1959). A general theory of measurement—Applications to utility. *Naval Research Logistics Quarterly, 6*, 283–294.

Quiggin, J. (1982). A theory of anticipated utility. *Journal of Economic Behavior and Organization, 3*, 324–343.

Rozeboom, W. W. (1962a). The untenability of Luce's principle. *Psychological Review, 69*, 542–547.

Rozeboom, W. W. (1962b). Comment. *Psychological Review, 69*, 552.

Savage, L. J. (1954). *The foundations of statistics.* New York: Wiley.

Shepard, R. N. (1978). On the status of 'direct' psychophysical measurement. In C. W. Savage (Ed.), *Minnesota studies in the philosophy of science,* Vol. IX (pp. 441–490). Minneapolis, MN: University of Minnesota Press.

Shepard, R. N. (1981). Psychological relations and psychophysical scales: On the status of "direct" psychophysical measurement. *Journal of Mathematical Psychology, 24,* 21–57.

Simon, H. A. (1955). A behavioral model of rational choice. *Quarterly Journal of Economics, 69,* 99–118.

Simon, H. A. (1978). Rationality as process and as product of thought. *The American Economic Review: Papers and Proceedings, 68,* 1–16.

Slovic, P., & Lichtenstein, S. (1968). Importance of variance preferences in gambling decisions. *Journal of Experimental Psychology, 78,* 646–654.

Stevens, S. S. (1946). On the theory of scales of measurement. *Science, 103,* 677–680.

Stevens, S. S. (1951). Mathematics, measurement and psychophysics. In S. S. Stevens (Ed.), *Handbook of experimental psychology* (pp. 1–49). New York: Wiley.

Stevens, S. S. (1957). On the psychophysical law. *Psychological Review, 64,* 153–181.

Stevens, S. S. (1959). Measurement, psychophysics, and utility. In C. W. Churchman & P. Ratoosh (Eds.), *Measurement: Definitions and theories* (pp. 18–63). New York: Wiley.

Suppes, P., Krantz, D. H., Luce, R. D,. & Tversky, A. (1989). *Foundations of measurement* (Vol. II). New York: Academic Press.

Suppes, P., & Winet, M. (1955). An axiomatization of utility based on the notion of utility differences. *Management Science, 1,* 259–270.

Tversky, A., & Kahneman, D. (unpublished). *Cumulative prospect theory: An analysis of decision under uncertainty,* manuscript.

von Neumann, J., & Morgenstern, O. (1947). *The theory of games and economic behavior* (2nd ed.). Princeton, NJ: Princeton University Press.

4 Chaos Theory: A Brief Tutorial and Discussion

James T. Townsend
Indiana University

PREFACE

The chapter begins with a quote from Sir Francis Bacon. This preface can do no better than another quote from that famous gentleman. It was a singular occasion where he departed somewhat from his customary modesty, the place being the Proem to "The Advancement of Learning," around 1600–1605.

> For myself, I found that I was fitted for nothing so well as for the study of Truth; as having a mind nimble and versatile enough to catch the resemblances of things (which is the chief point), and at the same time steady enough to fix and distinguish their subtler differences; as being gifted by nature with desire to seek, patience to doubt, fondness to meditate, slowness to assert, readiness to reconsider, carefulness to dispose and set in order; and as being a man that neither affects what is new nor admires what is old, and that hates every kind of imposture. So I thought my nature had a kind of familiarity and relationship with Truth.

This statement clothes William K. Estes like the proverbial glove. That nature has arguably led to some of the most valuable contributions to psychology in the 20th century. It has further guided several generations of experimental and theoretical psychologists through to their Ph.D. degrees and fruitful careers in research and teaching. There is little doubt that Kay C. Estes has been an inextricable part of the 'Estes equation' over much of their professional lives (see, e.g., Estes in Lindzey, 1989). Together they form a dynamic duo in psychology if there ever was one. For these reasons, the subsequent chapter is dedicated to both of them. May they continue their tradition of profound involvement and general invigoration of the science of psychology for many years to come.

"Those who aspire not to guess and divine, but to discover and to know . . . who propose to examine and dissect the nature of this very world itself, to go to facts themselves for everything." From "The Interpretation of Nature, or the Kingdom of Man", by Sir Francis Bacon, circa 1603.

This quotation from Francis Bacon emphasizes at once the preeminence of empiricism in the gaining of knowledge. It goes beyond that in hinting at the determinism to be found in observation of nature. One need not "guess" if one has gone to the root of the matter through direct inspection of natural causes. That kernel of deterministic causality has been an integral part of science now for hundreds of years. Even following the advent of mathematical probability theory in the 18th century, chance was long thought to be a manifestation of ignorance, rather than something inherent in the phenomena themselves. Quantum mechanics changed all that, but Albert Einstein, for one, refused to accept its finality of explanation, though admiring its predictive powers and consistency (e.g., see Pais, 1982, chap. 25).

With chaos theory, we have a new enigma and paradox: disorder, jumble, and confusion from certainty itself. It is apparent that fundamental issues arise in epistemology, philosophy of science, but also questions as to whether it can play a productive role in scientific theory and observation. This short essay can only touch the surface, but seeks to interest the reader in further rigorous study of the subject.

Chaos theory appears destined to be a major trend in the sciences. Perhaps it already is, to judge from its visibility in the serious specialized, the semi-serious, and the sheerly popular media. It is getting a rather slow start in the behavioral sciences, but signs abound that it is beginning to take off. When any such movement gets underway, there is some danger of faddism. If readers have any doubts that chaos theory, like many such "new items," can develop faddistic characteristics, I refer them to a recent article in one of the (usually) more responsible popular science magazines (McAuliffe, 1990).[1] The latter piece of scientific journalism saw, of course, chaos theory as being the new Nirvana, especially in biomedical and biopsychological spheres, but also managed to entangle deterministic chaotic processes with ordinary stochastic processes; in fact, it was entangled with just about anything that smacked of some degree of randomness.

Chaos theory would seem to be, on the face of it, made to order for psychology. It is deterministic and therefore seems to be strongly causal, yet often acts in a random, if not bizarre fashion. Is this not like human behavior? What is its real

---

[1]Considering the sorry state of science education and the dearth of graduating scientists in the United States, one can only applaud the efforts of popular science outlets to bring rather esoteric scientific information to the general public. However, there does seem to be a point at which the distortion may be so great that more harm than benefit is likely.

promise in psychology? How has it been used so far? How can it be employed to best advantage? Can we avoid the pitfalls that pandering to such vogues may augur?

One imposing obstacle in attempting answers to these and similar questions lies in the area of concern itself. Chaos theory is at base a highly technical, if not abstruse subject, embedded as it is in the field of nonlinear dynamics. Even physicists and engineers well versed in many aspects of dynamics have often found themselves unprepared for the kinds of mathematics required for the serious study of chaotic processes. Thus, the content of traditional courses in differential equations and systems theory simply did not cover those techniques. Indeed, even mathematics majors typically received only a modicum of the quantitative knowledge base pertinent to that end. Engineers and physicists frequently never took a course in real analysis or point-set typology—absolutely indispensable for deep understanding of this topic. Differential geometry and topology and algebraic topology form the foundation of the most general treatment of chaos theory and nonlinear dynamics. Chaos theory is now part and parcel of nonlinear dynamics and as such is seeing day-by-day advances in a myriad of fields.

On top of all this is the fact that there is no firm accepted definition of "chaos theory." There are several aspects that can be included in the definition or omitted, although some aspects seem more inherent than others.

This chapter provides the reader with an introduction to the main concepts of chaos theory. I attempt to strike a balance between the purely metaphorical or verbal presentation and a truly rigorous foundation. The former is often all that can be found in the popular literature, whereas to completely fulfill the latter might require a year or so of concentrated study. Several sources for additional inquiry are offered along the way. A good place to begin might be the historically and epistemologically exciting book by Gleick, "Chaos: Making a New Science" (1987) and the intriguing "Order out of Chaos" (1984) by Prigogine. The reader interested in a more rigorous background could time-share Gleick with a more technical introduction such as Devaney's "An Introduction to Chaotic Dynamic Systems" (1986) and/or Barnsley's "Fractals Everywhere" (1988). This pair can be rounded out with some strong visual intuition through the visual mathematics series on dynamics by Abraham and Shaw (1985) or computer simulations through Kocak's PHASER (1989), or Devaney and Keen's edited volume on chaos and fractals (1989). Each of these has received excellent reviews although not all mathematical reviewers have been ecstatic with the mathematical part of the history offered by Gleick, (not to mention a mild disparagement of psychology; see p. 165). Some types or regions of application, especially in psychology, are mentioned and commented on. Chaos theory is compared with certain other major trends that we have seen over the past few decades, and its ultimate implications for science and psychology considered.

CHAOS AS PART OF DYNAMICS

It is fascinating, and not always appreciated in the general scientific community, that chaos can arise in extremely simple dynamic systems. To make this a little more evident we need the notion of a difference or differential equation. Difference equations are appropriate in discrete time whereas differential equations apply in continuous time. Much of our discussion can be handled in discrete time. A set of difference equations representing N inputs and N outputs, and of the first order, can be written

$$x_1 (n + 1) = f_1 (x_1(n) , x_2(n) , \ldots , x_N(n) , n)$$

$$x_2 (n + 1) = f_2 (x_1(n) , x_2(n)) , \ldots x_N(n) , n)$$

$$\cdot \quad \cdot$$

$$\cdot \quad \cdot$$

$$\cdot \quad \cdot$$

$$x_N (n + 1) = f_N (x_1(n)), x_2(n)) , \ldots , x_N(n) , n)$$

Note that the state of each dependent variable, say (x_2), can in general be an arbitrary function of the states of all the dependent variables at time n as well as the time itself, n. The order of a difference equation stands for the number of differences across time (akin to the kth derivative in calculus) needed to determine the dynamics of a system. Alternatively, it says that at time n, the nth state of the system can depend on the states, moving k steps in time back from the present. A single input, single output system that is of order k can be generally written

$$x(n + k) = f(x(n + k - 1), x(n + k - 2), \ldots , x(n), n).$$

Systems with N input and N outputs or N inputs and M outputs, and of order $k > 1$, can be similarly constructed. The sequence of states of the system across time, n, is referred to as the "trajectory." In the latter case, with a single input and output, the trajectory is denoted by the variable x(n). In the former case, it is designated by the N-dimensional vector.

Often, the number of subsystems N (corresponding to the number of equations) is referred to as the dimension of the system. A single input, single output system thus has dimension 1. As noted, if we can write $x(n + 1) = f(x(n),n)$, then it is only of the first order. The fact that n is present as a separate independent variable in the function allows the evolution to be a direct function of time itself; for instance, the system might age, warm up, or change in some other fashion. The general expression f(x(n),n) also suggests that the state transition can depend on an input, so that the state of the system at time n + 1 depends not just on the state at time n but also on the current input. For instance, perhaps $y(n + 1) = f(x(n)) = \log(x(n)) + \sin(n)$. The second term on the right hand side represents a sinusoidal input function. The first term shows that the next state is also a nonlinear (in this case), logarithmic function of the current state. In such a case, we say that the system is being "forced" by the input function sin(n). If it is possible to write $x(n + 1) = f(x(n))$ alone, then we say we are dealing with the "free response" of the system, or that the system is "autonomous." The latter case is not only important in its own right (e.g., it covers such examples as a freely swinging pendulum) but it also turns out that the free response of a system is typically required to predict what happens when an input is added.

A further move toward simplicity takes place when we require that the function "f" be a linear operation: $f(x) = a(n)x$. Note that "a" can in general still directly depend on time n. If so, we have a "time dependent system." If not, we have a "time invariant system." In a time invariant system, the next state transition depends only on the current state and input. It cannot depend on time outside of that. For instance, this constraint rules out purely time related changes to a system, such as fatigue, warm-up effects, or the like, that are not explicit functions of the state and input.

Thus, we have $x(n + 1) = ax(n) + u(n)$ as the simplest linear system with input u. It is first order, linear and time invariant. If u is always equal to 0, we have an autonomous system. Now it happens that there are chaotic systems that require very little transformation of this type of system, that is, they possess only a little bit of "nonlinearity," and therefore enjoy some entitlement to the claim of simplicity. However, we shall see that their behavior can be quite complicated indeed.

Perhaps the most elemental chaotic system is the innocent looking discrete logistic system:

$$x(n + 1) = ax(n)[1 - x(n)] = ax(n) - ax^2(n)$$

Its states at time n, $x(n)$, can take on any positive or negative value, in general. However, the interesting behavior occurs on the interval $I = [0,1]$. Note that the only nonlinearity is the squaring of the state at time n. It is even time invariant, not to mention first order and one-dimensional! Even this elemental system exhibits many of the exotic facets of more complex, multidimensional systems. It has been much employed to represent population growth. Note for instance that if at time 0, $x(0)$ starts less than 1 but greater than 0, and the growth parameter "a" is greater than $1 + x(n)$, then growth occurs from time n to $n + 1$. For certain growth parameter values of "a," growth is smooth and approaches a fixed point, also known as an equilibrium point, as a limit. A fixed point is a state that remains unchanged under the dynamics of the system. Hence if a state is a fixed point, the system remains in that state forever. In this system, the fixed points are $x = 0$ and $x = (a - 1)/a$. Observe that only the second fixed point depends on the parameter "a." With two or more dependent variables, hence with dimension greater than or equal to 2, such systems can describe a wide variety of prey–predator interactions (e.g., see Beltrami, 1987).

My subsequent presentation owes much to Robert Devaney's excellent text, "Introduction to Chaotic Dynamical Systems" (1986). It focuses on discrete time systems, thereby avoiding some of the heavy topological machinery. Yet it includes analysis, in a quite rigorous fashion, of some of the best studied 1- and 2-dimensional chaotic systems. Of course, no text can do everything and this one leaves out discussion of the relationship of chaotic behavior to stochastic processes, for instance, ergodicity. A rigorous study of the latter requires some measure theory, which Devaney considered too far afield for this particular book. To be sure, this aspect is of interest to many groups, including psychologists, who might like to have a deterministic account of the randomness they regularly observe in human behavior. The present chapter contains some discussion of this side of chaotic activity.

We continue now with the logistic system already mentioned, $x(n + 1) = ax(n)[1 - x(n)]$. There is a quite nice way to plot successive states in a discrete system that illustrates exactly what is happening, at least in the short term. First we draw $y = f(x)$ as a function of x. This is, of course, the state transition function or map. The top of Fig. 4.1 shows the logistic function when $a = 3.41$

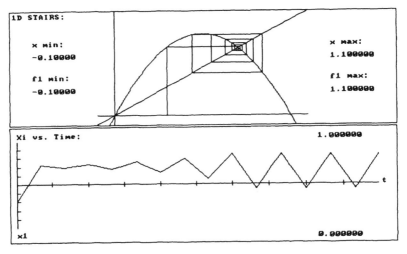

FIG. 4.1. The upper part shows a trajectory of the logistic system with growth parameter a = 3.41 and starting at x(0) = .30. The stair-step method of plotting the trajectory is used with t = 15, that is with 15 iterations. The lower part plots the evolution of the states as a direct function of time (iterations). The abscissa in the bottom graph is arbitrarily divided into 10 intervals, but the 15 interactions can be identified by the peaks and troughs. The ordinate in the bottom graph runs from 0 to 1.

and it is clearly quadratic in appearance. This and the other illustrations in this paper were plotted using PHASER (Kocak, 1989), a simulator and animator for elementary dynamic systems. It is inexpensive and worthwhile as a teaching device and for numerical investigation of nonlinear dynamic systems (see Townsend, 1991 for a review of PHASER).

Picking a starting point, in this case $x(0) = .3$, we draw a vertical line from the x-axis to where it intersects the function. This is, of course, the next value of the system state, that is, of $x(1) = .72$. We here round off to two decimals, as a matter of course. We next move horizontally over to the diagonal that represents the new value of $x(1)$ and prepares for the next transition to $x(2)$. The value of $x(2)$ is located by again drawing a vertical line, in this case down, to the function $f(x)$, and again over to the diagonal, and so on, yielding $x(2) = .68$. We refer to this type of graph as a "stair-step" function. This figure shows the first 15 iterations from our start point. It seems to be going into a cycle; that is, toward periodic behavior, and in fact that is what is occurring. This is supported by the bottom part of Fig. 4.1, which shows the time course of the states as a function of discrete time. Observe that the last few iterations are cycling back and forth between the same two values, indicating a period of 2. These values are approximately $x = .54$ (the point on the curve at the lower right hand corner of the biggest square) and $x = .78$ (the point on the curve at the upper right hand corner of the same square). In fact, with the same parameter $a = 3.41$, the identical period will be approached from almost any starting point (initial condition).

As already mentioned, there is no universally accepted definition of chaos. Some demand only the existence of a "strange attractor" (more on that later) whereas others require more. We proceed to follow Devaney's line of thought, which is topological, that is, oriented toward behavior of states of the system (considered as sets of points in a space) under the transformation defined by the difference equation. One may alternatively take a measure theoretic approach that facilitates connections with concepts of randomness. Before giving definitions, however, we do need a modicum of notation and concepts.

Let J be a space of points, which represent states of the system. For instance, J might be a vector space representing state of knowledge, emotional status, and other psychological characteristics, expressed numerically. Let f be the function that carries the state from time n to time $n + 1$. If f is applied k times, for instance to carry the state at time 1 to time $1 + k$, then we write $f^k[x(1)]$, where it should be noted that the superscript k is not a power, but rather the number of iterations of f. As an example, in the logistic system, $x(n + 1) = f[x(n)] = ax(n)[1 - x(n)]$, or expressed simply in terms of the position of the state, $f(x) = ax[1 - x]$.

The first criterion for chaos in our scheme is "topological transitivity," which says that if an observer looks at two sets of states, even if they do not overlap, the function f will always carry them together sometime in the future. More directly we meet the following definition.

Definition 1: Topological Transitivity and Dense Trajectories

A function f mapping the space J to J is said to be topologically transitive if for any pair of open sets U,V contained in J, there exists a positive integer k such that $f^k(U) \cap V \neq \phi$.#

The symbol ϕ represents the empty set. In order to be rigorous, we have retained the specification of open sets, which intuitively simply means one can always find another such set completely surrounding any point in the original set. An example of an open set is the open interval (0,1), which contains all numbers between 0 and 1 but not those endpoints themselves. All open sets composed of real numbers can be represented as unions of any number of such open intervals (a,b) and finite numbers of intersections of such unions. Expressly, Definition 1 indicates that iteration of the states represented by the set U always leads to at least one of them ending up in the set V. One outcome of this definition is that it is impossible to divide up the space into segregated regions that exhibit quite distinct kinds of behavior.

Another rather abstract notion we need is that of "density" of points in a certain set. A subset of points is dense in a set containing it, if any point in the larger set is arbitrarily close to a point in the original smaller set. For example, the set of rational numbers is dense in the set of all numbers. It turns out that in several chaotic systems of importance, obeying Definition 1 implies that there exists at least one trajectory through the space J that is dense in J. That is, there is an actual trajectory that sooner or later comes arbitrarily close to any point in the space! Of course, there may be, and usually will be, other non-overlapping trajectories in J that are not dense. In some ways, the possibility of a dense trajectory begins to give an impending sense of bizarreness—perhaps greater than the original, more general definition itself.

Let us take a new parameter value for "a" in the logistic system, a = 4, which is known to lead to chaotic behavior. Figures 4.2a, 4.2b, 4.2c show first 5 then 15 then 100 iterations, to give some idea of the evolution of the system, each time starting at the same state x(0) = .3. Both the stair-function as well as the time function as follows suggest that most areas of the interval from 0 to 1 are being visited when we allow time to become large.

The next definition is likely more familiar to readers who have exposed themselves to nontechnical but still objective articles appearing in *American Scientist, Science,* and other reputable general science magazines. It refers to a type of instability that arises in chaotic systems. Consider a system whose "initial state" we specify at time t = 0 and then observe as it evolves further over time. In chaotic systems, if the initial state is changed by any amount, no matter how small, the result in the future will be a series of states that are totally different from those associated with the first initial condition. This has come to be known as the "butterfly effect," the name suggested by meteorologist and early "chaotician" Edward Lorenz in a AAAS paper in 1979: "Predictability:

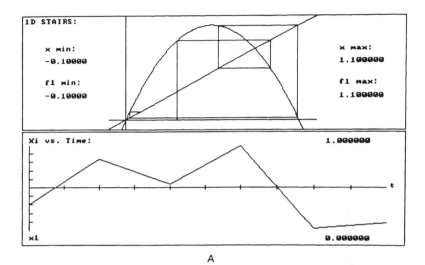

A

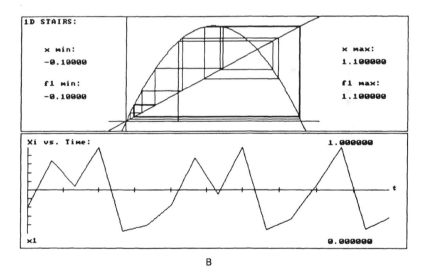

B

FIG. 4.2.   The logistic system is now illustrated along the same lines as Fig. 4.1, but now with a chaotic growth parameter a = 4. Again we start at x(0) = .30. Part A shows 5 iterations, Part B shows 15 iterations, and Part C shows 100 iterations.

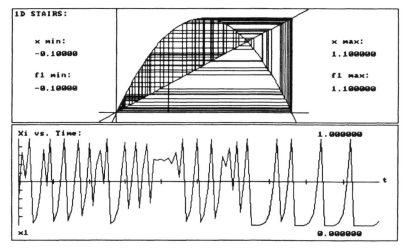

FIG. 4.2. (*Cont.*) C

Can the flap of a butterfly's wings in Brazil set off a tornado in Texas?" The idea is that the world might turn out very differently just according to whether the butterfly is flapping his or her respective wings (the initial condition) or not. The technical name is "sensitive dependence on initial conditions." In order to prepare for the definition, let "d" stand for a positive number. "N" will represent a small neighborhood (technically, this will be an open set; e.g., on the line of numbers, it might just be an open interval surrounding a point) around a state (i.e., point of J, called x). Finally let us represent the distance between the separate n-fold iterations of two points x and y, by $D[f^n(x), f^n(y)]$, where D is a metric imposed on the space J. Then we have

Definition 2: The Butterfly Effect

The function f, mapping J to J has sensitive dependence on initial conditions if there exists $d > 0$ such that for any x contained in J and any neighborhood N of x, there exists a y contained in N also, as well as an $n \geq 0$ such that $D[f^n(x), f^n(y)] > d.\#$

We can translate this to mean that there is some distance greater than zero, such that no matter where we start out (i.e., x), and no matter how little we perturb it (i.e., x to some y within an arbitrarily small distance from x), if we go far enough into the future (i.e., at least n iterations), we will always find that x and y have evolved to states that are at least distance d apart. It should be noted that not all y within that small distance of x need to have trajectories that separate from x's trajectory in the future, but there must exist at least one such y for any small neighborhood of x. Figure 4.3, computed with PHASER (Kocak, 1989) for a = 4 and two different starting points, x(0) = .300 and y(0) = .301, exhibits the butterfly effect because after only 8 iterations, the states of the two systems are far apart, despite their initial closeness. From then on, the two trajectories bear little similarity to one another.

One upshot of this critical aspect of chaotic behavior is that even a tiny error in measurement can lead to a vast misprediction of future behavior. There are also critical implications for computers, but these have to be bypassed in this treatment (but see, e.g., McCauley, 1986, chap. 5; or Grassberger, 1986). The "sensitive dependence" axiom is virtually always present, explicitly or implicitly, in discussions of chaos. Sometimes topological transitivity is missing, and sometimes the following characteristic is also missing, despite its profound importance for dynamic behavior.

We all know what periodic behavior is in a system: States continually repeat themselves as the system evolves. Those with sensory science backgrounds are familiar with Fourier theory and its importance in describing signals, their transmission, and decoding. Simple sine waves are the purest form of auditory stimulation. One of the primary regular types of behavior of well-behaved systems is periodic. A variation on this theme is when the amplitude of the periodic function is modified, for instance, damped, where the amplitude of the (perhaps complex)

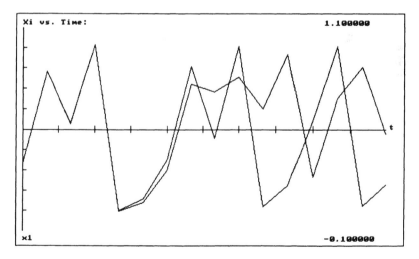

FIG. 4.3.   Illustration of the Butterfly Effect in the logistic system with a = 4. Two start values x(0) = .300 and y(0) = .301 are so close together that they cannot be discriminated on the scale used in the figure, yet within about eight iterations their graphs deviate substantially and shortly will be totally unalike.

periodic waves goes toward zero. Unstable, but otherwise easily describable linear systems can evidence periodic behavior altered by the amplitude swinging more and more wildly out of control.

It turns out that periodic functions can emerge in chaotic systems, too, but often in a rather untoward fashion. We again require the notion of "density" of sets of points. In our case, the points are still to be thought of as states of the system. Following Devaney (1986), we nevertheless refer to this aspect as an "element of regularity."

Definition 3: Dense Periodic Trajectories

A system is said to possess an element of regularity if there exists a dense subset of periodic points in the set of states J. That is, for any point x contained in J, there is a point y that continually iterates back to itself, that is, arbitrarily close to the original point x.#

This infinite set of periodic trajectories exists for a single parameter value, say "a" in the logistic system, whenever the system is chaotic. Moreover, it has been shown that for any value of "a," there exists at most one trajectory that is attracting, in the sense that any initial state (within some set of initial states) will asymptotically approach that trajectory. Such an attracting trajectory is also known as an attracting "orbit." Some authors prefer "orbit" to "trajectory"; we hereafter use them interchangeably. When an attracting orbit or trajectory is periodic, it is referred to as a "limit cycle." In any event, it follows that at most one of the infinite set of periodic orbits in a chaotic logistic system is attracting! All the rest are repelling. This in turn implies that only one of them—the attracting trajectory—is ever likely to be observed in nature, as the probability of landing exactly on the periodic orbit to start with is zero. In fact, some chaotic systems may have no attracting periodic orbits whatsoever. This is the case in the logistic system when a = 4, as we learn in more detail as follows.

In Fig. 4.1, "a" (a = 3.41) was selected so that there exists just one periodic orbit in toto. This system is not chaotic. For "a" greater than 1 and less than 3, there exists a single equilibrium point and no periodic orbits. At a = 3, we see our first periodic orbit, of period 2, and this continues to hold until a = 3.449499 (to the sixth decimal place) at which value the system gains a period 4 limit cycle (attracting trajectory or orbit). Thus, our value of a = 3.41 is still (barely) within that range and does not yield chaotic behavior. As "a" continues to be increased it more and more frequently (i.e., with successively smaller increments required) hits values where more and more *even* periodic orbits appear. That is, these periods are all powers of two; this behavior on the part of "a" is referred to as a "period doubling sequence." In fact, period doubling occurs for a sequence of "a" where $a_k = a^* - c/F^k$, F = 4.669202, c = 2.637 and $a^* = 3.569946$ is the limit of this sequence. (These numbers are approximations to the unending decimal true values.) The constant "F" is known as the "Feigenbaum constant" due to its discovery by Mitchell Feigenbaum in 1975. Observe that here F^k is actually the kth power of F, and also that each of these periods associated with a particular value of "a" is an attracting orbit. When a $>$ a*, we first encounter chaos and a $= 4$ is usually considered the most chaotic system within this logistic family, one reason being that the system is chaotic on the entire interval I $=$ [0,1]. Other interesting aspects of the system at a $= 4$ are mentioned later and note that Figs. 4.2, 4.3, and 4.5 were plotted with that parameter value. The reader is referred to Lauwerier (1986) for more detail on these phenomena in the logistic and other systems.

The three foregoing definitions establish a foundation for chaotic behavior: (a) sets of states, denoting, for example, distinct initial conditions, always eventually march into one another's province, (b) no matter how close two initial conditions are, they become unpredictably separated in the future, and (c) there are infinite numbers of periodic states as close as one likes to any other state in the system.

ALLIED NOTIONS OF CHAOS

There are a number of other concepts that are intimately related to chaotic activity. Under certain conditions, they follow from the previously described conditions, but may also themselves be used as part of the defining qualities of chaos.

Fractals, Cantor Sets, and Fractal Dimension

Loosely speaking, fractals are sets of points that are "self-similar" under magnification. An intuitive example would be a Christmas tree each of whose branches was a replica of the original tree, each of its needles was yet another replica containing its own "branches and needles," and so on. Additional sources for reading about fractals are Mandelbrot, 1983, and Barnsley, 1988. Mandelbrot has been the prophet, if not the godfather, of fractals for many years.

The mathematician Georg Cantor, in the course of establishing an axiomatic basis for the number system and the differential and integral calculus in the 19th century, invented what naturally came to be known as "Cantor sets." The most classic version starts out as the set of all numbers between zero and one. Then the (open) middle third of points is removed; that is, the points lying between $\frac{1}{3}$ and $\frac{2}{3}$, but not including the latter. Next, the middle thirds between $\frac{1}{9}$ and $\frac{2}{9}$ and between $\frac{7}{9}$ and $\frac{8}{9}$, but not including the endpoints, are removed and this continues forever. Finally, the Cantor set is all the remaining points left. This is formally constructed by taking the set intersection of the first remaining set with the second remaining set, this with the third, and so on. Figure 4.4 shows the first three removals of open intervals of powers of 3 (i.e., remove the middle $\frac{1}{3}$, then

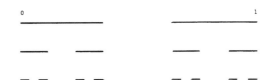

FIG. 4.4. The construction of the "missing thirds" Cantor set is illustrated by the first three removals, of the middle third, the next middle ninths, and the next middle twenty-sevenths. The blank parts are those that have been deleted. The Cantor set is what is left after the unending sequence of continued deletions.

remove the two middle $\frac{1}{9}$'s of the remaining intervals, then the four middle $\frac{1}{27}$'s of the remaining intervals and so on).

The resulting set of points apparently rendered many mathematicians of the late nineteenth century apoplectic, due to its strange characteristics. For instance, it contains not only an infinite set of points but an uncountably infinite set; in mathematics terms, that is as many points as there are numbers on the line to start with! Yet taking the usual measure of sets of points finds that this set has measure equal to zero. In a probabilistic sense, there would be probability zero of ever landing on one of them. A related property is that the set contains no intervals; that is, no subsets of points that can be described as "all the numbers lying between the endpoint numbers "a" and "b." It is thus said to be "totally disconnected." It is furthermore "perfect," by which is meant that a region around any point in the set, however small that region, always contains other points from that set as well. The self-similarity is apparent on "blowing up" any section between, say, 0 and $(\frac{1}{3})^n$. Such a section, on magnification by 3^n, looks exactly like the original set.

Even though such odd mental creatures have been accepted by most mathematicians for the greater part of the 20th century, it is only with the increasing prominence of nonlinear dynamics and chaos that they have begun to enjoy respectability and daily employment in science and technology. Benjamin Mandelbrot (1983) is largely responsible for making the study of fractals almost a science in itself (despite his publicized antipathy toward scientific specialization), including suggestions of how nature often may employ fractal structure.

We nevertheless must be cautious in separating the use of fractals in describing static architecture, as in the structure of the brachiating alveoli in lung tissue as opposed to their presence in chaotic dynamics. It was formerly thought that the branching took place in an exponential manner, but recent evidence supports the fractal model. Fractals play a very important role in actual dynamics, too.

We now take a moment to establish the main areas of chaos of the logistic system, $x(n + 1) = ax(n)[1 - x(n)]$. We want to concentrate on the dynamics starting in the interval [0,1], henceforth to be called "I." Obviously, for any value of "a," x = 0 is an equilibrium, or fixed point, because if the system is on that point it never leaves. When a > 1, the more interesting equilibrium point x* = (a − 1)/a does lie in I, and in fact the most intriguing dynamic occurs for even larger values of the growth parameter "a." Thus, for any value of "a" between 1 and 3, starting at any point x(0) in I leads inevitably toward the equilibrium point x*. The equilibrium point I is therefore in this case asymptotically stable as all other points converge toward it. However, this is definitely not chaos.

When a > 3 exciting things begin to happen, as mentioned earlier. We haven't space to develop matters in detail. But, recall that the number of periodic points increases as "a" gets larger until at a = a* = 3.5699 approximately (note that this a* is the same one mentioned earlier), one reaches an attracting periodic orbit with an infinite period! Now, as we advance beyond a = a* toward a = 4, we encounter small windows of "a" where there exist attracting periodic orbits. Interspersed among these windows chaos reigns.

At a = 3.839 approximately, and within a nonchaotic window, there are an infinite number of periodic trajectories, but only one of these, with period three, is attracting. And, it is feasible to compute how many points are associated with any particular period. For small intervals around any of the three points of the attracting orbit, the trajectory of any point in such an interval becomes ever closer to the periodic trajectory and thus mimics it. This attracting periodic trajectory constitutes an instance of a limit cycle, which is, as defined earlier, a relatively simple type of behavior that nevertheless cannot be found in linear dynamic systems. (Figure 4.1 revealed the existence of a limit cycle in our earlier discussion.) All other sets of periodic points for this value of "a" are repelling. Hence, within small intervals around such points that do not overlap the small intervals just mentioned, trajectories starting from within those intervals are repelled away from the periodic trajectory in question, and toward the attracting orbit of period three.

Now let us jump to a > 2 + $\sqrt{5}$ where we have true chaotic behavior in the sense of Definitions 1, 2, 3. Observe in beginning that here a > a*. Moreover, the chaotic behavior all occurs on a Cantor set contained in I. Any point outside that Cantor set is repelled to −(infinity), so the chaotic set of points in I is repelling. Indeed, it is possible to show that the numbers in I that are eventually mapped out of this interval, and into the negative real line, correspond exactly to the open intervals removed by the construction of the Cantor set previously presented. What is left is the chaotic (Cantor) set. Correspondingly, if the initial condition x(0) is within the Cantor set, then chaotic behavior is guaranteed.

There may be, and probably is, some profound yet compelling reason for the appearance of this highly regular, if bizarre set of points, outside the manifest production of it by the dynamics. However, so far I have not discovered any completely satisfying statement to this effect. Interestingly, one may achieve a bit more sense of the topology of construction of such Cantor sets by going on to view certain two-dimensional chaotic systems such as the horseshoe map (e.g., Devaney, 1986). Such an exercise would go beyond the present scope, but readers are encouraged to avail themselves of the opportunity to do so.

An additional note of interest here is that although the point of equilibrium x* is definitely unstable in that no matter how close you start to it, points may be found there that wander off, the system satisfies "structural stability," a concept due apparently to the Soviet mathematicians Andronov and Pontriagin around 1935 (see Jackson, 1989). The concept has become an important tool in modern investigations of dynamics through the work of Stephen Smale and others (Smale, 1969). Structural stability means, roughly, that a small change in the parameters of the system lead to small changes in the overall behavior of the system: the qualitative actions are the same. So a small change to "a" around the value $2 + \sqrt{5}$ leaves the system chaotic.

A quite special value of the parameter is found at a = 4, the value employed before in Figs. 4.2 and 4.3. Here, the system does not obey structural stability, so changing "a" just a little to greater than or less than 4 provides drastic alterations in the system's behavior. This system, with a = 4, is actually chaotic on the entire interval I.

A strategic related concept is that of "fractal dimension" (see, e.g., Barnsley, 1988; Mandelbrot, 1983). We are all familiar with the usual idea of dimension as given, say, by the classical D-dimensional Euclidean space, and represented pictorially by D othogonal coordinates. It turns out that we need a refined notion to handle the "size" of fractal sets of points; for instance, states in the phase space of system trajectories. Thus, although the usual measure of the Cantor set is 0, this odd subspace has a non-zero fractal dimension. There are many generalizations of the ordinary notion of dimensions. The following has become fairly standard in fractal arithmetic.

For most of our cases of interest, we can think of our space J of points (i.e., states of the system) as being closed and bounded subsets of an ordinary Euclidean space. Because of its lack of intervals or analogous sets in higher dimensional Euclidean spaces, the dimension of J can only be approximated in a finite sense. Again let J stand for our space (i.e., set of points) and D stand for the fractal dimension, whatever it turns out to be. Suppose that $N[(\frac{1}{3})^n]$ is the smallest number of abstract boxes, each of dimension D, each side of which is of length $(\frac{1}{3})^n$ that will cover J. That is, N is the number of such boxes, but N depends on the size of the individual boxes, which in turn depend on the lengths of each side, namely $(\frac{1}{3})^n$. The particular fraction $\frac{1}{3}$ with which we represent the declining box sizes is arbitrary; here it makes the computation of dimension of the Cantor set particularly straightforward.

To paraphrase this discussion, we can loosely think of the "volume" of J as being given by these N "boxes," each of size $[(\frac{1}{3})^n]^D$. Observe that this is akin to the ordinary box size in three dimensional space with sides of length $(\frac{1}{3})^n$ being $[(\frac{1}{3})^n]^3$. That is D = 3 in the standard case. So in effect we are putting together N boxes, each with sides of length $(\frac{1}{3})^n$ units and dimension D (and therefore they are D-dimensional boxes) to make up the entire volume. Thus, we have $N[(\frac{1}{3})^n] \times [(\frac{1}{3})^n]^D = V$, the approximate volume. However, it is only an approximation, albeit an increasingly accurate one, as n becomes larger, and therefore the individual boxes become smaller.

Now let $V = 1$, which is simply setting an arbitrary but convenient unit of volume, and take natural logarithms of both sides to obtain approximately

$$\log\{N[(\tfrac{1}{3})^n]\} + \log[(\tfrac{1}{3})^{nD}] \doteq 0$$

Performing some algebra we find that

$$D \doteq \frac{-\text{Log}\{N[(\tfrac{1}{3})^n]\}}{\text{Log}[(\tfrac{1}{3})^n]} = \frac{\text{Log}\{N[(\tfrac{1}{3})^n]\}}{\text{Log}(3^n)}$$

Finally, taking the limit as n grows large yields

$$D = \underset{n\to\infty}{\text{Lim}} \frac{\text{Log}\{N[(\tfrac{1}{3})^n]\}}{\text{Log}(3^n)}$$

Applying this formula to the Cantor set finds that $D = .6309$, thus delivering a fractional or "fractal" dimension.

One way of discovering chaotic behavior is to measure the dimension and learn that it is fractional (and therefore fractal). Of course, as is the case with most of the means of pinpointing chaos, sufficient precision must be maintained that the investigator can be reasonably certain that the fraction part of the dimension is not simply measurement error or other "noise."

Lyapunov Exponents, Randomness, and Strange Attractors

Another valuable index of dynamic activity is that of the so-called "Lyapunov exponent" (see, e.g., McCaulley, 1986; Wolf, 1988). Lyapunov, a Soviet, was an early innovator in dynamics and, along with others such as Henri Poincare in France and G. D. Birkoff in the United States, was responsible for many central concepts in nonlinear systems theory. Although the Lyapunov exponent is one of those concepts that is likely to seem rather abstruse, in fact we can readily gain some pretty good intuitions about it. It is related to several critical aspects of chaotic behavior, but perhaps most immediately to the butterfly effect, that is, "small differences in initial state lead to big differences later on." Recall that the implication is that a tiny measurement error of the initial state of a system means that future predictions may be vastly wrong.

We take a general discrete difference equation as our starting point, $x(n + 1) = f[x(n)]$. Now suppose that the error in specifying the state $x(n)$ at time n, is $dx(n)$. That is, the true state is $x(n)$ but due to some type of measurement error we believe it to be $x(n) + dx(n)$. It then follows that $x(n + 1) + dx(n + 1) = f[x(n)$

+ dx(n)]; or written in terms of the error at time $n + 1$, $dx(n + 1) = f[x(n) + dx(n)] - x(n + 1)$. That is, the error at time $n + 1$ must be the difference between what the true state should be and the propagated error from the previous time epoch. The next step is to take a linear approximation, using Taylor's series from elementary calculus to get

$$dx(n + 1) = dx(n) \cdot f'[x(n)],$$

where f' denotes the derivative or rate of change of the function with respect to the state x at time n. But this is the simplest type of linear difference equation, easily solved to obtain, in absolute value of error,

$$|dx(n)| = |dx(0)| \cdot |f'[x(0)] \cdot f'[x(1)] \dots f'[x(n - 1)]|.$$

Thus, the error is seen to depend on the way in which f' changes with system state and time. For a large class of systems it happens that this error either grows exponentially fast (the chaotic systems) or decays exponentially fast (the well-behaved systems). This in turn implies that $dx(n) = dx(0) \cdot 2^{Vn}$, where V is the Lyapunov exponent. But these two equations may be set equal to one another and solved for V yielding

$$V = (1/n) \sum_{i=0}^{n-1} \text{Log}_2\{f'[x(i)]\}$$

That is, V is approximately equal to the arithmetic mean of the logarithms of the derivative of f. Moreover, in the limit this operation precisely defines the Lyapunov exponent to be

$$V = \underset{n\to\infty}{\text{Lim}} \left\{ (1/n) \sum_{i=0}^{n-1} \text{Log}_2\{f'[x(i)]\} \right\}$$

A positive value for V implies the butterfly characteristic associated with chaos, and means that two trajectories beginning very close together can move apart very rapidly. Incidentally, it can be shown that V measures the rate at which information is lost from the system over time.

Now for the "random" aspect of chaos. In some ways, this is one of the most fascinating topics, because chance seems to emerge from determinism, an antinomy. However, most readers of this chapter have, perhaps unwittingly, used just this mechanism in the form of computer random number generators, which are deterministic means of producing numbers with some probability distribution. Even a thorough intuitive discussion would require considerably more space than we have available, but perhaps a few comments can motivate the idea.

First, observe that in any deterministic system of the kind almost always used to describe real-life mechanisms, the trajectory is unique given a starting point. Therefore any possible trajectory has probability 1 or 0 depending on the particu-

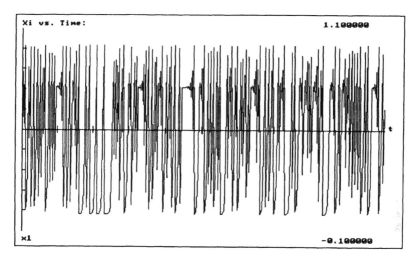

FIG. 4.5.   The random appearance of the logistic system with a chaotic growth parameter a = 4 is shown with 400 iterations and starting at x(0) = .30.

lar start point that is selected. No probability yet. Furthermore, in the absence of periodicity, single points or states will have probability 0 of ever being visited twice, so just how should we look for chance in such situations?

A key lies in the earlier notion of topological transitivity. Recall that this assumption implies that for any open set (no matter how small) in the state space, there will be points in any other open set of states (also no matter how small that second set), such that starting at one of those points will lead in a finite time to a visit to the first open set; that is, to one of the states contained in that set. An implication is that if we are experimentally measuring the relative frequency of visiting a particular set of states, there will always be a finite measure of frequency of visitation to that set. Furthermore, the relative frequency of visitation can be taken as a measure of probability of finding the state to be contained within that set. Figure 4.5 gives some indication of randomness on the part of the logistic system when its growth parameter is in its chaotic domain. The ability of the sequence of states to span the entire available range for this parameter value (i.e., from 0 to 1), is apparent in the figure. However, the frequent and seemingly random visiting of midranges can also be appreciated from the "thicknesses" of the lines in the center parts of the graph.

In some very important cases, both physically and mathematically, the trajectory will be "randomly mixing" as well as "ergodic." Randomly mixing means that all sets of states of equal measure are visited equally often by a trajectory. A second, closely related interpretation is that if one generates a large number of trajectories simultaneously, such as placing a droplet of ink into a jar of water, each molecule of water corresponding to a separate state, then ultimately any volume of the state space will contain an amount of the original trajectories in proportion to its percentage of the total volume.

To be "ergodic" means that time averages approximate and eventually approach the space averages. This is a concept that is also useful in ordinary probability and stochastic process theory (e.g., Papoulis, 1984), but has special relevance here for deterministic processes. One intriguing sidelight is that random mixing implies ergodicity. In order to obtain an intuitive grasp of this consider any one-dimensional difference equation $x(n + 1) = f[x(n)]$, operating on a state space of size S. Observe the time average $\bar{x}(n) = (1/n) [x(1) + x(2) +...+ x(n)]$. Now segregate the visited states $x(i)$ into N bins of equal size, say, of size s. And let $n(i)$ be the number of times (iterations) that $bin(i)$ is visited. Also, let $\hat{x}(i)$ stand for the approximate value of x within $bin(i)$. This permits us to rewrite $\bar{x}(n) = (1/n) [n(1) \hat{x}(1) + n(2) \hat{x}(2) +...+n(N) \hat{x}(N)]$. This is only the usual way that psychologists classify data when taking averages and performing other computations.

Now though, we can see that random mixing implies that visits to the ith bin occur in the ratio $n(i)/n = s/S = \Delta S$; that is, the frequency of visits to any bin is proportional to its relative size, and all of these were constructed to be equal. Therefore we can further express our average as $\bar{x}(n) = [\hat{x}(1) \Delta S + \hat{x}(2)\Delta S +...+ \hat{x}(N)\Delta S]$, and we have turned the time average into a space average, which must approach the real average by the law of large numbers. Also, if we let the bin sizes ΔS decrease toward infinitesimals, dS, then the average becomes an integral instead of a sum, $\bar{x} = \int \hat{x} \, dS$. Note too that dS takes on the properties of a density function in this case.

There are a number of other concepts that are important features of nonlinear and especially chaotic dynamics that unfortunately cannot be dealt with here. One that must be touched on because of its ubiquity in the popular press is the "strange attractor." At this point in time, there is no firmly agreed-on definition of this concept, but we now limn in something of what is available. First of all, as employed earlier in the chapter, an "attractor" of any kind is a set of points representing states of the system that trajectories approach. Typically an attractor forms an "invariant set" in the sense that if a trajectory starts within the set, it can never leave. In the special case of a single isolated point-attractor, we have the classical "asymptotically stable" state that is also an equilibrium point on account of its being an invariant set.

When an attracting set of points is itself periodic, this means that the approaching trajectories act more and more like the periodic set itself as time increases. This is the by now familiar concept of a "limit cycle," behavior that although not too complex, is never found in linear dynamic systems as noted earlier. Some authors define strange attractors by exclusion, as any attractor that is neither an ordinary asymptotically stable equilibrium point nor a limit cycle.

Usually a candidate for the status of a strange attractor is some version of a fractal set, although some writers have proselytized for yet more exotic criteria (some aspects of this debate are offered in Mandelbrot, 1983).

Interestingly, the earlier logistic system, with parameter "a" > $[2 + \sqrt{5}]$, possesses a fractal (a Cantor) set as an "invariant repellor," rather than an attractor. That is, once in the Cantor set, the trajectory can never depart, but if a trajectory is outside the repelling set, it marches away. Such a set of states thus merits the name "strange repellor." On the other hand, a number of well studied chaotic systems do possess fractal sets as attractors. And a full characterization of the attractors of some systems, though much scrutinized, are still not well understood. The Hénon map is a case in point (e.g., Devaney, 1986), and is stated as the beguilingly simple two-dimensional discrete system

$$x(n + 1) = a - by(n) - x^2(n)$$

$$y(n + 1) = x(n).$$

So far, even the employment of computer "experiments" to suggest regularities that are then proven rigorously has not fully broken the code of this system.

CHAOS, SCIENCE, PHILOSOPHY, AND PSYCHOLOGY

When even such trite-looking systems as the logistic or Hénon can be so appallingly difficult to analyze, is there any hope for a science of chaotics? The answer seems to be a tentative "yes." Already, as can perhaps be anticipated from our earlier discussion, a quite organized set of concepts, methods, and classes of behavior has evolved. Chaos theory is deeply embedded in general dynamic systems theory, its mathematical roots going back to the early 20th century. Some mathematicians actively dislike the term "chaos" because it seems to connote total disorder and an inability to ever be analyzed, understood, or dealt with. This is far from the truth as great progress has been made in understanding the phenomena, and mathematical control theory can in some cases be applied to "dechaotize" a chaotically behaving system.

Further, many scientists would claim that nonlinear dynamics in general and chaotics in particular has substantially benefitted progress in physics, chemistry, and engineering. Turbulence is a good example where chaotic dynamics has played a major role in advancing the field. Chaos theory seems to give a much better description of transitions to "noisy" behavior than does the time-honored Reynolds approach. Nevertheless, an improved ability to encompass spatial, rather than just temporal patterns in turbulence would be welcomed (e.g., Pool, 1989d).

As one would expect, there are spheres of research where the science and philosophy interact and sometimes almost merge. An instance is the study of "quantum chaos," a domain characterized by much contention and debate (see e.g., Pool, 1989c). One provocative if over-simplified question: Can chaos theory provide the answer to Einstein's long-lived disgruntlement with the absence of a deterministic or at least a causal underpinning to the probabilities of quantum theory (see e.g., Born, 1971; Pais, 1982, chap. 25)? So far, no acceptable deterministic infratheory to quantum mechanics has survived the theoretical and experimental crucible (but see the keen discussion by Suppes, in press-a).

A major enigma is that the very substance of chaos emerges from the trajectories, so far a meaningless construct in quantum mechanics. Nevertheless, there are those who seem to see quantum theory as a kind of parallel probabilistic model of chaos. For instance, the physicist Jensen (cited in pool, 1989c) suggests that instead of attempting to define "quantum chaos," it would be more profitable to determine the characteristics of a quantum system that correspond to chaos in the corresponding classical system. Others seem to feel that quantum activities should evince their own uniquely "quantum" type of chaos. In any event, presently the two theories, although agreeing on a number of predictions, diverge on several important dimensions. Thus, quantum behavior can be viewed, in most instances, as reversible (e.g., Feynman, 1965). This is a possibility unknown in chaos where the initial state is lost forever under evolution of the system.[2] Such questions could clearly be of strategic importance to future developments in quantum theory, not to mention the sheer intellectual attraction they have.

Outside of the physical sciences, the "dishes are not yet done." Certainly biology has already profited from the influx of nonlinear dynamics for many years (see, e.g., Glass & Mackey, 1988), yet the verity and usefulness of chaos theory are far from accepted by all physiologists and physicians (e.g., see Pool, 1989b). Now, there is little dispute that certain types of biological structure earlier thought to be of more mundane architecture (e.g., the aforementioned brachiation of alveoli tissue in the lung), is better described by fractal patterns, than, say, by exponential branching.

However, when one looks for indubitable evidence for chaos in the actual dynamics, it can be elusive. Even in population biology and meteorology, fields

---

[2]The only exception of which I am aware, of particle physics, where time is asymmetric, involves the decay of the neutral K meson. Also strictly speaking, a deterministic chaotic system is usually taken as evolving in one direction. However, because of the butterfly effect, one would have to know the initial state to an infinitesimal degree of accuracy in order to retrace the time sequence, impossible in reality. Another related point is that the mathematical form, sans physics, of Newtonian trajectories is typically time-symmetric: Differential equations are mute with regard to whether time runs from the left to right or vice versa. An engaging nontechnical discussion of the history and modern conceptions of physical time is "Arrows of Time" by Richard Morris (1984). Finally it is worth noting that the "arrow" offered by the second law of thermodynamics in terms of increasing entropy is based on statistical laws applied to closed systems. Hence, the time-symmetric character of the laws of individual particles is not violated.

boasting early pioneers in chaos research, respectively Robert May and Edward Lorenz, there are a number of respected scientists who downplay the evidence and importance of chaotic models (cf. Pool, 1989a,d).

This is true in physiology, again despite the contribution being made with nonlinear dynamics, and as forthrightly admitted by Glass and Mackey in their intriguing book "From Clocks to Chaos: The Rhythms of Life" (1988), and appears also in such fields as epidemiology (Pool, 1989a). One trend in recent years, especially salient in biological circles, is to view chaos as a "positive" rather than "negative" dynamic. Thus, it is claimed that in a number of organic systems, the heart being one of them, chaos is the preferred healthy state (e.g., Goldberger & Rigney, 1988). Others maintain that the normal brain evidences chaotic activity whereas epileptic seizures are associated with an overabundance of periodicity (e.g., Rapp, Zimmerman, Albano, de Guzman, Greenbaun, & Bashore, 1986). Here too, there is yet controversy (Glass & Mackey, 1988).

Perhaps the most frequent use of chaos theory so far in the behavioral sciences is in a directly "philosophy of science" fashion to warn of the indeterminacies of psychology. To be sure, such admonitions are requisite for any discipline studying phenomena that are reasonably complex. In fact, Suppes (1985), among others, has invoked chaos theory in dealing with such enigmas as the determinism versus probabilistic dichotomy. Are all natural phenomena, no matter how seemingly random they may appear, simply evidence of complex deterministic mechanisms? One "classic" example discussed by Suppes (1991) is that of a special-case three-body problem where any random sequence of coin tosses turns out to represent a solution to the dynamics of this "simple," and in Newtonian terms, decidedly deterministic, special case. The orbit of Pluto, impossible to predict in the long run, is another Newtonian example of chaos. As Suppes observes in another paper (1985), prediction in more complex arenas, such as real-world political scenarios, is likely to remain an "after-the-fact" reconstruction rather than true a priori rigorous anticipation. Suppes has suggested that the dichotomy between stability and instability might be a more appropriate partition than determinism versus indeterminism, given the theoretical and empirical results associated with chaos.[3]

Within psychology, Uttal (1990) has recently put forth a tractate listing cogent

---

[3]Although generally agreeing with Suppes' various points, I do suggest caution in dealing with the notion of "stability," because more than one definition is possible. Typically, stability can be defined in constrained but simple terms for linear systems (e.g., "nonpositive real parts of eigenvalues," "bounded inputs produce bounded outputs," and the like). Then for more general systems, new definitions often encompass the more elementary ones. For instance, Lyapunov stability is implied in linear systems by the two definitions already given. Nevertheless, in general contexts, even when the state-space is metric and the systems are autonomous (that is, they are evolving without input), a number of useful types of stability (e.g., Lyapunov, Lagrange, Poisson, Structural) can be advanced that are not necessarily equivalent (see, e.g., Jackson, 1989; Sibirsky, 1975). These diverse notions of stability may be of both philosophical as well as scientific value.

reasons that theorizing, especially reductionistic modeling, in psychology may be critically limited in its possibilities. The reasons are, briefly and not in the author's order, chaos, second law of thermodynamics, combinatorial complexity, black box identifiability problems; and, with much less emphasis, models are only abstractions of a full reality; the cumulative microcosm does not equal the macrocosm, mathematics is descriptive, not reductive, and Goedel's theorem. In my opinion, certain of the objections, such as the limits on system identification in automata (e.g., Moore, 1956), can be viewed as "the glass is half full" rather than "half empty." That is, Moore's and related theorems show that it is generally possible, from finite inputs and finite outputs, to determine the system down to a canonical class of systems within which no further identification may be possible. Given the complexity of systems and the relative poverty of input–output data, the ability to resolve the potential models down to a constrained subclass could be viewed as quite impressive.[4]

Within more standard information processing theory and methodology in psychology, the evolution of experimental technology in deciding the serial versus parallel processing issue can be interpreted in a similar optimistic way (Townsend, 1990). First, the early work, extending from the 19th century (Donders, 1869) to Sternberg's crucial experiments (1966, 1969) one hundred years later, provided much needed data, increasing experimental and theoretical sophistication and a motivating force to parallel–serial research. The second phase found that the standard experimental paradigms were primarily either testing other aspects of processing (e.g., limited vs. unlimited capacity) or quite restricted examples of the classes of models under investigation (Townsend, 1971, 1972, 1974). The third phase, overlapping with the second and still in progress, has discovered fundamental distinctions between parallel and serial processing, a number of which can be used to experimentally test the opposing concepts (Snodgrass & Townsend, 1980; Townsend, 1972, 1974, 1976a, 1976b, 1984, 1990; Townsend & Ashby, 1978, 1983).

To be sure, the case Uttal introduced for ceilings on understanding and prediction due to chaos was compelling. Nevertheless, it may be that a large portion of psychological laws might be developed that are able to skirt the tiger traps of chaos. For instance, whether random behavior emanates from chaos or pure indeterminism, averages and other statistics can be employed to reveal lawful if stochastic, for example, correlations, power spectra, and so on, systemic behavior. Perhaps an apposite example is that of reaction time. Certainly most reaction time experiments have been in the context of relatively simple cognitive operations. Yet no investigator has even found the same reaction time given on two successive trials, no matter how simple the mental task. Reaction time has been an extraordinarily useful tool for investigators of elementary perceptual and

---

[4]Uttal (personal communication) reminds me that he is primarily designating the limits on reductionism per se, that is, on our ability to discover the exact material entities and functions producing our behavior, rather than on modeling as a descriptive tool.

cognitive processes (e.g., Luce, 1986; Posner, 1978; Townsend & Ashby, 1983; Townsend & Kadlec, 1990; Welford, 1980). Lawfulness has emerged even though it may not be understood for a long time just where the randomness of reaction time derives, in the physiological sense; or whether the randomness is deterministic in its origin.

Another example, apt both for applied physics, engineering, and chemistry, as well as for sensory psychology, is the fact that linear systems can account so well for first order descriptions of phenomena. Much of engineering sciences is based on linear approximations to reality that have built bridges, houses—and instruments of war!—for hundreds of years, not to mention electronic equipment from around the turn of the century to the present. It is primarily with the advent of such fields as nonlinear optics (e.g., lasers, diffraction), superconductivity, and the need for more efficient air foils and turbines that nonlinear dynamics has forced itself on the engineering community. In vision and acoustic research, linearity suffices to give a very good first approximation to the functioning of the systems, although certain nonlinear phenomena have been known for some time.

Interestingly, the dependent variable "accuracy" has, alongside reaction time, been a mainstay in cognitive research. And accuracy is most frequently described in terms of probability correct; again, the phenomena and models are (sometimes implicitly) assumed to be inherently random. The origin of the probabilism is not necessarily critical for the erection of laws governing behavior, even for other indices of behavior such as rating scales, EEG recordings, and so on.

What about more direct uses of chaos, where it plays an important part in a psychological theory? There is not too much to report here so far, although Killeen (1989) argues cogently for an important post for dynamics in general, in describing human behavior. Probably the most intense development to date is the dynamic theory of olfactory discrimination advanced by Walter Freeman and his colleagues (Freeman, 1987; Freeman & Di Prisco, 1986; Freeman & Skarda, 1987). Backed up by several decades of electrophysiological data and an evolving dynamic theory of elementary olfactory processes, chaos serves a strategic purpose in the theory. Chaos is present in the state space at all phases of respiration, but not in all parts of the state space. For odor discrimination, learning leads to formation of limit cycles that attract learned odors during early inhalation. If an odor fails to "find" an attracting cycle, it falls into a chaotic well and is then identified as "novel." Further, chaotic activity is seen as ". . . a way of exercising neurons that is guaranteed not to lead to cyclic entrainment or to spatially structured activity. It also allows rapid and unbiased access to every limit cycle attractor on every inhalation, so that the entire repertoire of learned discriminanda is available to the animal at all times for instantaneous access." The last quote comes from an open commentary paper in *Behavior and Brain Sciences* (1987) by Freeman and Skarda, to which the reader is referred for an introductory account of the theory, a thorough and lively debate with the commentators, and a healthy set of references.

Notably, not all scientists believe chaos is necessary to explain such phenomena. Grossberg is one of these (see, e.g., Grossberg's commentary in the cited *Behavior and Brain Sciences* article), who has developed his own account of numerous psychological and psychobiological behaviors (Grossberg, 1982). Another intriguing exploration of nonlinear dynamics in the context of sensation psychology is found in the recent book by Gregson (1988). Here chaos does not play as integral a role as in Freeman's scheme, but may appear under various circumstances (see also, the in-depth review of Gregson's book by Heath, 1991). It remains to be seen whether chaos in fact is a strategic linchpin in psychophysical discrimination, but it appears that the Freeman and colleagues' corpus of results constitute the most impressive theoretical venture in psychobiology employing chaos to date.

My friends in the "hard" sciences and mathematics assure me that the behavioral sciences are not alone in the experience of fads and trends. However, we certainly do seem to have our share of them. I was tempted to substitute "suffering" in place of "experience" in the first sentence of this paragraph, but caught myself. When I thought of some of the various trends in experimental psychology over the past 50 years or so, it became apparent that although most of them were a bit "oversold," each in turn eventually took its place in the overall armamentarium of the science. And each added something pretty unique and helpful in doing scientific psychology.

In order to put some limits on things, suppose we start back around 1913, the time of the advent of behaviorism and, in this country, of Gestalt psychology. Later came neobehaviorism, then in the late 1940s and 1950s the trends associated with cybernetics, logic nets, game theory, automata, signal detection, and information theory. The latter, especially, were undoubtedly critical in leading inexorably to the information processing approach and artificial intelligence culminating in the field of cognitive science. But intersticed into the early 1950s was the birth of mathematical psychology as a subfield of psychology.

Some would say mathematical psychology was a fad. Mathematics goes back to the roots of experimental psychology in the 19th century, especially in psychophysics. However, Ebbinghaus and others used simple functions to describe the progress of learning as well relatively early. William Estes pioneered in the more formal development of the field with his seminal "Toward a statistical theory of learning" in 1950. This article presaged the forthcoming "stimulus sampling theory" that continues to play an important role in memory and learning theory today. The Estes paper was followed closely by a Robert Bush and Herbert Mostellar paper in 1951, introducing their linear operator approach. The early 1950s also saw the emergence of the influential signal detectability theory by the Michigan group composed of psychologists and electrical engineers (see Tanner & Swets, 1954, and their references). Another valuable cog in the mathematical machine was formal measurement theory (also called axiomatic measurement theory, foundational measurement theory, the representation approach, and so on), initiated in psychology by Patrick Suppes (1951), Duncan Luce (1956),

Clyde Coombs (1950). These investigators and their colleagues began to formalize S. S. Stevens' (1946) informal schema of scale types varying in strength, and the use by Louis Thurstone (1927) and Louis Guttman (1944) of ordinal data to constrain possible psychological scales. The desire to further develop the notions of von Neuman and Morgenstern's (1944) utility theory approach to decision making was also present in some directions of this early work.

Multidimensional scaling has been another trend that stemmed from the work of investigators like Coombs and Guttman but also to some extent from a more psychometric tradition (e.g., factor analysis, and the metric scaling efforts of Young & Householder, 1938). Later under the impact of contributions by Shepard (1966) and Kruskal (1965), multidimensional scaling took on a separate life of its own and has occasionally been abused. But, when employed in a careful way, particularly in the context of a substantive theory, it has also contributed to our understanding of many phenomena.

In the ensuing years, mathematical psychology took up the cudgel in almost every area of psychology including the still active information processing approach (Atkinson & Shiffrin, 1968; Estes & Taylor, 1964; Laberge, 1962; Link & Heath, 1975; Townsend, 1971, 1972). Although mathematical psychology is not, if it ever was, viewed as the panacea to all psychology's dilemmas, it has apparently become an inseparable part of the psychological panorama. Even the Markov approach, highly voguish in the 1960s and thereafter falling into decline, continues to enrich theory in learning and memory (e.g., Riefer & Batchelder, 1988).

Artificial intelligence held sway as the dominant force in cognitive science from the early 1960s until recently, with some of the major players in the field being Feigenbaum, Minsky, Papert, Suppes, Newall, Simon, and McCarthy. John Anderson and Gordon Bower have been prolific contributors to cognitive science, in particular through their synthesis of artificial intelligence and mathematical modeling strategies.

The Johnny-come-lately, but with perhaps the broadest and greatest early inertia, is connectionism, although it has its roots in work going back many years. Its hallmark is its connotations and occasionally denotations of neural and biological realism. Its mathematical underpinnings are largely in dynamic systems theory, although stochastic processes and other aspects of applied physics play a role. Earlier pioneers were Rashevsky, Ashby, and McCulloch and Pitts. This new field intersects many disciplines, including electric engineering, computer sciences, artificial intelligence, neurosciences, psychology, and philosophy. Yet certain of the seminal resurgence papers have been published in the *Journal of Mathematical Psychology* (e.g., Grossberg, 1969; Hoffman, 1966) or other theoretical psychology journals (e.g., Anderson, 1973; McClelland, 1979), and several leaders of the movement, namely, Rumelhart and James Anderson, received doctoral or postdoctoral training in mathematical psychology laboratories.

Undoubtedly, there will be a lot of "noise" associated with this, as with the other vogues, but, just as surely, the approach will offer something that has been missing in our theoretical battery. In the present instance, in my opinion, what has been lacking in most cognitive science theory oriented around the metaphor of digital computers (or automata in general) has been the sense of a naturally functioning dynamic system. That has been more important than the "neuralism" per se. Connectionism has also upped the ante in forcing cognitive scientists to go one level deeper in their assumptions about the "process" aspects of their model; deeper, say, than the level reached by the typical "production system" routines.

Now along comes chaos theory, although as mentioned earlier, it really has a history going back to the late 19th and early 20th century. We have seen that outside some relatively narrow areas of the physical sciences, it has still not captured universal approval, despite its big splash. I suspect chaotic mechanisms will be increasingly visible in psychological models. Much of its form will be metaphorical rather than mathematical. When it assumes a mathematical formulation, it will still likely be difficult to come up with data that is sufficiently tight relative to the difficulty in estimating chaotic parameters, to test the chaotic part of the model (see, e.g., chapters by Wolf and by Grassberger in Holden, 1986; plus other references in Glass & Mackey, 1988). In some cases, its contribution may be by way of giving a cogent deterministic interpretation to an observably random phenomenon.

The metaphorical uses are not necessarily to be sneered at. The very sensationalism of chaos theory has attracted the attention of many who might not otherwise have been drawn to consider dynamic systems theory as a workable tool. Thus, it can help areas or problems previously conceived of in a static fashion to move towards a more dynamic, and therefore perhaps psychologically and biologically realistic accounting of the data. This seems to be happening in developmental psychology, for example, long treated by most psychologists as a series of static plateaus, with little concern about how the evolving individual gets from one stage to the next.

In addition, chaos will almost certainly be adopted by connectionism, because as intimated, virtually all nonlinear as well as linear dynamics can be considered "connectionistic" in some broad sense; the only necessary requirement being the allusion to neuralistic conceptions.

So far its greatest contribution has been to challenge mathematicians and other scientists to dig deeply into the general and profound aspects of nonlinear dynamics.[5] This, in turn, may offer the field of psychology a more powerful technology with which to study complex cognitive systems and behavior.

---

[5] A recent integrative and readable presentation that places chaos nicely within the subjects of dynamics and neutral phenomena is "Exploring Complexity" by Nicolis and Prigogine (1989).

ACKNOWLEDGMENTS

The impetus for this chapter comes from my own longstanding interest in dynamics and systems theory (e.g., Townsend and Evans, 1983). Its specific theme emerged from a talk I gave at the 1990 Interdisciplinary Conference at Jackson Hole. I am indebted to Stephen Link, Richard Shiffrin, and William Uttal for their insightful comments on a draft of this manuscript. I thank Joseph Stampfli for inviting me to participate in his and John Ewing's excellent seminar on chaos theory and for his participation in my own seminar on nonlinear systems and dynamics theory. The writing of this chapter was partly supported by NSF Grant #8710163 from the Human Cognition and Perception Section.

REFERENCES

Abraham, R. H., & Shaw, C. D. (1985). *Dynamics—The geometry of behavior*, Parts 1 through 4. In The Visual Mathematics Library, R. H. Abraham (Ed.). Santa Cruz, Ca: Aerial Press.

Anderson, J. A. (1973). A theory for the recognition of items from short memorized lists. *Psychological Review, 80,* 417–438.

Atkinson, R. C., & Shiffrin, R. M. (1968). Human memory: A proposed system and its control processes. In K. W. & J. T. Spence (Eds.), *The psychology of learning and motivation: Advances in research and theory, Vol. 2.* New York: Harper & Row.

Barnsley, M. (1988). *Fractals everywhere.* San Diego: Academic Press.

Beltrami, E. (1987). *Mathematics for dynamic modeling.* New York: Academic Press.

Born, M. (1971). *The Born-Einstein letters: The correspondence between Albert Einstein and Max and Hedwig Born, 1916–1955.* New York: Walker & Co.

Bush, R. R., & Mosteller, F. (1951). A mathematical model for simple learning. *Psychological Review, 58,* 313–323.

Coombs, C. H. (1950). Psychological scaling without a unit of measurement. *Psychological Review, 57,* 145–158.

Devaney, R. (1986). *An introduction to chaotic dynamical systems.* New York: Addison Wesley.

Devaney, R., & Keen, L. (Eds.). (1989). Chaos and fractals: The mathematics behind the computer graphics. *Proceedings of the Symposium in Applied Mathematics,* Vol. 39. Providence, RI: American Mathematical Society.

Donders, R. C. (1869). On the speed of mental processes (W. G. Koster, Trans.). In W. G. Koster (Ed.), *Attention and performance, Vol. 2,* Amsterdam: North Holland.

Estes, William K. (1989). In G. Lindzey (Ed.). *A history of psychology in autobiography, Vol. VIII.* Stanford, CA: Stanford University Press.

Estes, W. K., & Taylor, H. A. (1964). A detection method and probabilistic models for assessing information processing from brief visual displays. *Proceedings of the National Academy of Sciences, 52,* 446–454.

Feynman, R. (1965). *The character of physical law.* Cambridge, MA: MIT Press.

Freeman, W. J. (1987). Simulation of chaotic EEG patterns with a dynamic model of the olfactory system. *Biological Cybernetics, 56,* 139–150.

Freeman, W. J., & Skarda, C. A. (1987). How brains make chaos in order to make sense of the world. *Behavioral and Brain Sciences, 10,* 161–195.

Freeman, W. J., & Viana Di Prisco, G. (1986). EEG spatial pattern differences with discriminated odors manifest chaotic and limit cycle attractors in olfactory bulb of rabbits. In G. Palm (Ed.), *Brain theory.* New York: Springer-Verlag.

Glass, L., & Mackey, M. C. (1988). *From clocks to chaos: The rhythms of life*. Princeton, NJ: Princeton University Press.

Gleick, J. (1987). *Chaos: Making a new science*. New York: Viking Penguin.

Goldberger, A. L., & Rigney, D. R. (1988). Sudden death is not chaos. In J. A. S. Kelso, M. F. Schlesinger, & A. J. Mandell (Eds.), *Dynamic patterns in complex systems ( )*. Singapore: World Scientific.

Grassberger, P. (1986). Estimating the fractal dimensions and entropies of strange attractors. In A. V. Holden (Ed.), *Chaos*. Princeton, NJ: Princeton University Press.

Gregson, R. (1988). *Nonlinear psychophysical dynamics*. Hillsdale, NJ: Lawrence Erlbaum Associates.

Grossberg, S. (1969). Embedding fields: A theory of learning with physiological implications. *Journal of Mathematical Psychology, 6*, 209–239.

Grossberg, S. (1982). *Studies of mind and brain*. Dordrecht, Holland: D. Reidel.

Guttman, L. (1944). A basis for scaling qualitative data. *American Sociological Review, 7*, 139–150.

Hoffman, W. C. (1966). The Lie algebra of visual perception. *Journal of Mathematical Psychology, 3*, 65-98.

Jackson, E. A. (1989). *Perspectives of nonlinear dynamics*. Cambridge: Cambridge University Press.

Killeen, P. R. (1989). Behavior as a trajectory through a field of attractors. In J. R. Brink & C. R. Haden (Eds.), *The computer and the brain: Perspectives on human and artificial intelligence ( )*. Amsterdam: North-Holland.

Koçak, H. (1989). *Differential and difference equations through computer experiments*. New York: Springer Verlag.

Kruskal, J. B. (1965). Analysis of factorial experiments by estimating monotone transformations of the data. *Journal of the Royal Statistical Society, Series B., 27*, 251–163.

LaBerge, D. (1962). A recruitment theory of simple behavior. *Psychometrika, 27*, 375–396.

Lauwerier, H. A. (1986). One-dimensional iterative maps. In A. V. Holden (Ed.), *Chaos*. Princeton, NJ: Princeton University Press.

Link, S., & Heath, R. A. (1975). A sequential theory of psychological discrimination. *Psychometrika, 40*, 77–105.

Lorenz, E. (1979). Predictability: *Does the flap of a butterfly's wings in Brazil set off a tornado in Texas?* Address at the Annual Meeting of the American Association for the Advancement of Science, Washington, D.C.

Luce, R. D. (1956). Semi-orders and a theory of utility discrimination. *Econometrica, 24*, 178–191.

Luce, R. D. (1986). *Response times*. New York: Oxford University Press.

Mandelbrot, B. B. (1983). *The fractal geometry of nature*. San Francisco: W. H. Freeman.

McAuliffe, K. (1990, February). Get smart: Controlling chaos. *Omni*.

McCauley, J. L. (1988). An introduction to nonlinear dynamics and chaos theory. *Physica Scripta*, Series T, Vol. T20.

McClelland, J. L. (1979). On the time relations of mental processes: An examination of systems of processes in cascade. *Psychological Review, 86*, 287–330.

Moore, E. F. (1956). Gedanken-experiments on sequential machines. In C. E. Shannon & J. McCarthy (Eds.), *Automata studies ( )*. Princeton, NJ: Princeton University Press.

Morris, R. (1984). *Times arrows: Scientific attitudes toward time*. New York: Simon & Schuster.

Nicolis, G., & Prigogine, I. (1989). *Exploring complexity*. New York: W. H. Freeman.

Pais, A. (1982). *'Subtle is the Lord. . .': The science and life of Albert Einstein*. New York: Oxford University Press.

Papoulis, A. (1984). *Probability, random variables, and stochastic processes*. New York: McGraw Hill.

Pool, R. (1989a). Ecologists flirt with chaos. *Science, 243*(Jan.), 310–313.

Pool, R. (1989b). Is it healthy to be chaotic? *Science, 143*(Feb.), 604–607.

Pool, R. (1989c). Quantum chaos: Enigma wrapped in a mystery. *Science, 243*(Feb.), 893–895.

Pool, R. (1989d). Is something strange about the weather? *Science, 243*(Mar.), 1290–1293.

Posner, M. I. (1978). *Chronometric explorations of mind*. Hillsdale, NJ: Lawrence Erlbaum Associates.

Prigogine, I. (1984). *Order out of chaos*. New York: Bantam Books.

Rapp, P. E., Zimmerman, I. D., Albano, A. M., de Gusman, G. C., Greenbaun, M. N., & Bashore, T. R. (1986). Experimental studies of chaotic neural behavior: Cellular activity and electroencephalographic signals. In H. G. Othmer (Ed.), *Nonlinear oscillations in biology and chemistry ()*. Berlin: Springer-Verlag.

Riefer, D. M., & Batchelder, W. H. (1988). Multinomial modeling and the measurement of cognitive processes. *Psychological Review, 95*, 318–339.

Shepard, R. N. (1966). Metric structures in ordinal data. *Journal of Mathematical Psychology, 3*, 287–315.

Sibirsky, K. S. (1975). *Introduction to topological dynamics* (L. F. Boron, Trans.). Leyden, Holland: Noordhoff.

Smale, S. (1969). What is global analysis? *American Mathematics Monthly, 76*, 4–9.

Snodgrass, J. G., & Townsend, J. T. (1980). Comparing parallel and serial: Theory and implementation. *Journal of Experimental Psychology: Human Perception and Performance, 6*, 330–354.

Sternberg, S. (1966). High-speed scanning in human memory. *Science, 153*, 652–654.

Sternberg, S. (1969). The discovery of processing stages: Extensions of Donders' method. In W. G. Koster (Ed.), *Attention and Performance, Vol. 2*. Amsterdam: North Holland.

Stevens, S. S. (1946). On the theory of scales of measurement. *Science, 103*, 677–680.

Suppes, P. (1951). A set of independent axioms for extensive quantities. *Portugaliae Mathematica, 10*, 163–172.

Suppes, P. (1985). Explaining the unpredictable. *Erkenntnis, 22*, 187–195.

Suppes, P. (1990). Probabilistic causality in quantum mechanics. *Journal of Statistical Planning and Inference, 25*, 293–302.

Suppes, P. (1991). *Indeterminism or instability, does it matter?* In G. G. Brittan, Jr. (Ed.), Causality, Method and Modality (pp. 5–22). Kluwer Academic Publishers.

Tanner, W. P., & Swets, J. A. (1954). A decision-making theory of visual detection. *Psychological Review, 61*, 401-409.

Thurstone, L. L. (1927). A law of comparative judgement. *Psychological Review, 34*, 273–286.

Townsend, J. T. (1971). A note on the identifiability of parallel and serial processes. *Perception & Psychophysics, 10*, 161–163.

Townsend, J. T. (1972). Some results concerning the identifiability of parallel and serial processes. *British Journal of Mathematical and Statistical Psychology, 25*, 168–199.

Townsend, J. T. (1974). Issues and models concerning the processing of a finite number of inputs. In B. H. Kantowitz (Ed.), *Human information processing: Tutorials in performance and cognition*. Hillsdale, NJ: Lawrence Erlbaum Associates.

Townsend, J. T. (1976a). A stochastic theory of matching processes. *Journal of Mathematical Psychology, 14*, 1–52.

Townsend, J. T. (1976b). Serial and within-stage independent parallel model equivalence on the minimum processing time. *Journal of Mathematical Psychology, 14*, 219–238.

Townsend, J. T. (1984). Uncovering mental processes with factorial experiments. *Journal of Mathematical Psychology, 28*, 363–400.

Townsend, J. T. (1990). Serial versus parallel processing: Sometimes they look like Tweedledum and Tweedledee but they can (and should) be distinguished. *Psychological Science, 1*, 46–54.

Townsend, J. T., & Ashby, F. G. (1978). Methods of modeling capacity in simple processing systems. In J. Castellan and F. Restle (Eds.), *Cognitive Theory, Vol. 3 ()*. Hillsdale, NJ: Lawrence Erlbaum Associates.

Townsend, J. T., & Ashby, F. G. (1983). *The stochastic modeling of elementary psychological processes*. Cambridge: Cambridge University Press.

Townsend, J. T., & Evans, R. (1983). A systems approach to parallel-serial testability and visual feature processing. In H. G. Geissler (Ed.), *Modern issues in perception*. Berlin: VEB Deutscher Verlag der Wissenschaften.

Townsend, J. T., & Kadlec, H. (1990). Psychology and mathematics. In R. E. Mickens (Ed.), *Mathematics and Science ()*. Singapore: World Scientific.

Uttal, W. (1990). On some two-way barriers between models and mechanisms. *Perception & Psychophysics, 48*, 188–203.

von Neumann, J., & Morgenstern, O. (1944). *Theory of games and economic behavior*. Princeton, NJ: Princeton University Press.

Welford, A. T. (Ed.). (1980). *Reaction times*. New York: Academic Press.

Wolf, A. (1986). Quantifying chaos with Lyapunov exponents. In A. V. Holden (Ed.),*Chaos ()*. Princeton, NJ: Princeton University Press.

Young, G., & Householder, A. S. (1938). Discussion of a set of points in terms of their mutual distances. *Psychometrika, 3*, 19–22.

5 Imitatio Estes: Stimulus Sampling Origins of Weber's Law

Stephen Link
McMaster University

INTRODUCTION

Error is a cause of variability in measurements. This Gaussian principle is so fundamental to psychological research that even a tremor of disbelief threatens to topple the vast edifice of statistical methods balanced so precariously upon it. In spite of such an unmistakable danger, perhaps it is time to question the traditional account of variability in psychological measurements. Among other things, this challenge to orthodox theory will provide a different view of the source of variability and, as an example of its utility, produce a new foundation for, and interpretation of, Weber's Law and Weber's Constant.

A student's first encounter with Gaussian theory is likely to occur during a scientific study of natural phenomena. Given the task of creating a concise description of a sample of clearly different observations, the student quickly learns that a single measure characteristic of the sample provides a very handy summary, especially when many different sets of data must be considered. The student's teacher, having revealed the advantages of using simple statistics, often trots from the stable of descriptive measures a few prime examples of well-behaved statistics. Their exemplary behaviors are evaluated in terms of the Theory of Error introduced by Gauss (1809, 1821).

Gaussian Theory

Gauss' theory[1] proposed that the true value of a measured object was shrouded in additive error. Specifically the *Observed* measure, *O,* depended on two compo-

---

[1] Surprisingly few students are taught that the basis for statistical analysis is a mathematical model that may or may not apply to their data. Rarely are they taught how to test the model that they so often

nents. The first component was the *True,* but unknown, value of the phenomenon. The second was an unknown amount of *Error.* Gauss proposed that the Observed measure resulted from adding together these two components:

$$O_i = T + E_i.$$ (1)

where i indicates the i-th observation, T is a constant, and E_i is the random error that is an additive part of the i-th observation. The Gaussian Theory of Error postulated that the average error equaled zero. Therefore, the average of the observations will equal the unknown but desired True value, T. This simple model of variability in the observed measures is so appealing that students often overlook the subtle change in thinking that the theory imposes on their thought. Variability is now a difficulty with measurement rather than a part of nature.

Nineteenth-century theorists proposed several different measures of variability. Laplace suggested the average absolute difference from the mean. Gauss suggested the *probable error* as a measure of variability. The probable error determined a distance of size d from the mean of a Gaussian distribution that included 25% of the distribution. Deviations from the mean greater than d, that is greater than one probable error, occurred with probability 0.25. The probable error, d, is related to the standard deviation, σ, currently a more widely used measure of variability, by the similarity transformation, $d = \sigma/0.674$.

Fechner's Theory

In the first bold effort to turn Psychology away from philosophical analysis and toward the rigors of scientific measurement, G. T. Fechner (1860/1966) extended the Gaussian theory to the comparison of sensations. Fechner assumed that a stimulus, such as a lifted weight, produced a sensation that varied due to an unknown amount of error added to the sensation's True value. In the case of a lifted weight, the sensation of heaviness varied because the sensory system combined a variable amount of error with the True heaviness to produce a variable sensation of heaviness. A weight failed to feel the same from one lifting to another because heaviness, like any other measurement, was burdened by the fickle influence of erratic error. Going one step further, Fechner supposed that this variability in sensation caused errors in judging which of two weights weighed more.

Fechner's extension of Gauss' theory postulated that the average heaviness of the two weights served as a decision criterion for deciding which of the two weights felt heavier. If the heaviness of a weight exceeded this threshold then the subject judged that weight to be the heavier of the two. The theory is illustrated in Fig. 5.1. The two Gaussian probability distributions show how much error is added to the True heaviness evoked by two physically different weights. The

---

apply, and, except for a very few courses about mathematical models, students seldom learn how to create models for their own unique situations.

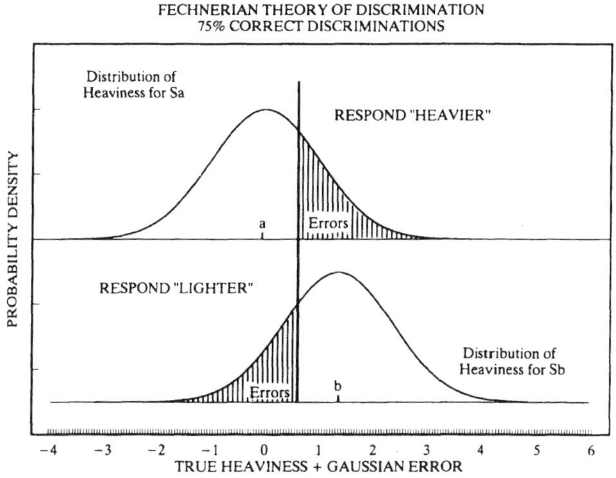

FECHNERIAN THEORY OF DISCRIMINATION
75% CORRECT DISCRIMINATIONS

FIG. 5.1.   Fechner's model for discriminating between two stimuli S_A and S_B.

ideal case shown in Fig. 5.1 illustrates 75% correct responding, and therefore 25% error, regardless of which weight is judged. In this case the decision criterion corresponds to one probable error and represents a 75% threshold for difference discrimination.

Fechner's genius was to observe that if the increase in sensation from a True heaviness value to the decision criterion equaled the value, d, and the difference between the two physical stimulus values equaled D, then d = D/2. The value of D is known but the value of d is unknown. Fechner argued that the Gaussian Theory of Error provided a method for determining the unknown value of d. In the case of 75% threshold responding, 25% of the sensations lie beyond the decision criterion and produce errors of judgment, instances where the subject experiences the lighter weight to be heavier, or vice versa. Fechner realized that this particular case occurred when the decision criterion was exactly 1 probable error above the mean.

This measure of the position of the decision criterion provided an indirect measure of the unknown amount of error imposed upon True heaviness by the sensory system. The decision criterion equalled two different values, one in terms of sensory system error, the other in terms of physical stimuli. Equating these two quantities gave

$$d = D/2.$$

Or, because $d = \sigma/0.674$ in standard deviation units,

$$\sigma = D(0.674)/2. \tag{2}$$

In this way the error in the sensory system, measured by d or σ, became observable. Differences between sensations became measurable. Psychology became a science.

Thurstone's Theory

The measurement of sensory error sparked much of the new Psychology's future progress. However, by 1927 L. L. Thurstone proposed an alternative to the Gaussian model. Variability, Thurstone argued, came not from error but from the application of the stimulus itself. In his famous article "Psychophysical Analysis," Thurstone (1927a) argued:

> I shall call the psychological values of psychophysics *discriminal processes*. . . . fluctuation among the discriminal processes for a uniform repeated stimulus will be designated the *discriminal dispersion*. . . . The discriminal dispersion which any given repeated stimulus produces on the psychological continuum is usually normal [Gaussian]. . . . Psychological measurement depends, then, on the adoption of [dispersion] as a base and the use of its standard deviation as a unit of measurement for the psychological continuum. (pp. 20, 22, 24)

Thereafter much of the analysis of stimuli depended upon Thurstone's creation of the Psychological Continuum.

For Thurstone, dispersion was an inherent feature of nature and society. An individual's attitude may show no variability and yield the same response to repeated questioning. Thus, within the individual no variability in attitude existed. Yet attitude varied within the society of individuals. Attitude could be measured because a comparative question requiring an attitude for a response evoked different responses from individuals who conveyed the dispersion of an attitude within the society. In answering an attitude question, a proportion of individuals prefers one choice response to another because the strength of their attitude is greatest for that choice.

Although Thurstone extended Fechner's theory, the new theory still allowed for dispersions within an individual. In "A law of comparative judgment" Thurstone (1927b) showed how the new theory of choice was consistent with both Fechner's Theory and Weber's Law. Moreover, the Law of Comparative Judgment accounted for newly observed relations among response proportions for comparisons between any two of a number of physical stimuli. The theory opened the door for Guilford (1928, 1931) to analyze paired comparisons of lifted weights and to scale the weights on a Psychological Continuum of heaviness.

Using a set of seven weights ranging to 215 g from 185 g in 5 g increments, Guilford constructed a full paired comparison design for judging the heaviness of lifted weights. Then, by using 100 observations per pair, a total of 4900 judgments, Guilford estimated a subject's probability of judging which of two weights was heavier. The obtained response proportions appear in Fig. 5.2. The

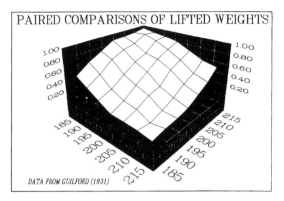

FIG. 5.2. Guilford's (1931) paired comparisons of lifted weights generated response proportions ranging from 0.01 to 0.99.

proportion of "heavier" judgments varies from a low of 0.01 for judging a 185 g weight to be heavier than a 215 g weight, to a maximum of 0.99 for judging a 215 g weight to be heavier than a 185 g weight.

Applying the Law of Comparative Judgment to the 7×7 matrix of response proportions netted Guilford a scaling of the weights on a Psychological Continuum of heaviness. By inverse application of the Law of Comparative Judgment, the psychological scale values for the weights generated a close reproduction of the 49 observed response proportions. Although Guilford did not determine how well the model fit the data, the seven estimated heaviness values predict response proportions that do deviate substantially from observed values. The chi-squared goodness of fit statistic is quite large, 36.3 based on $[(7 \times 7 - 7)/2 + 7] - 7 = 21$ degrees of freedom. Nevertheless, the seven numbers that defined heaviness on the Psychological Continuum[2] became summary statistics for the subject's 49 response proportions.

Perhaps Thurstone's greatest achievement was the creation of the Psychological Continuum. The idea is all the more remarkable because it seems so obvious. One must wonder why Fechner did not create it himself, or why nearly 70 years elapsed between Fechner's formulation of the theory of discrimination and the birth of the Psychological Continuum. Perhaps the Psychological Continuum was not so obvious after all.

This question prompted a brief discussion in Bill Estes' office at Harvard University (c. 1980). Estes' famous contribution to discrimination theory comes from Stimulus Sampling Theory (cf. Neimark & Estes, 1967). In the time-honored tradition of Leibnitz and Herbart, Stimulus Sampling Theory postulates that discrete elements act as theoretical units relating a response to a stimulus. An element may be connected to only one response, but sampling a set of such

---

[2]Paul Horst, interviewed by the author in 1983, credits Thurstone with the invention of the Psychological Continuum.

theoretical elements provides a certain proportion that are "connected" to a particular response. Stimulus Sampling Theory proposes that the proportion of sampled elements connected to a response equals the probability that the response occurs. Plenty of response variability occurs due to the random sampling of elements, but no Psychological Continuum exists at all! I asked Bill what he thought of using a Psychological Continuum and why one didn't appear in his own work. He replied, "Well, Steve, they're alright if you believe in them."

WAVE THEORY

Perhaps the Psychological Continuum and the units of dispersion used to measure distances between stimuli presumed to lie on the continuum were unnecessary. Perhaps a great deal of psychophysical analysis could be replaced by an approach more in keeping with Stimulus Sampling Theory and the urn models of early quantum mechanics. My first theoretical goal was to provide a new foundation for Weber's Law and Weber's Constant that bypassed the Psychological Continuum. The result proved to be quite satisfactory, generating Weber's Law, Weber's Constant, and Fechner's Law (Link, 1990, 1992) by way of a realizable physical process.

A startling discovery followed: The new theory required only a single parameter to predict the 49 paired comparison proportions from Guilford's classic experiment. That single parameter proved to be Weber's Constant. By knowing the value of Weber's Constant for lifted weights the response proportions were predictable in advance of the experiment. The general features of this new theory, Wave Theory, are introduced in the following. A more extensive treatment is available elsewhere (Link, 1992).

Stimulus Representation

Judgments about stimuli are comparative. Comparisons are often made between stimuli presented to the external senses such as the heaviness of two weights, the brightnesses of two lights, the tastes of two vegetables. Such stimuli are the generators of quantum pulses of electrical energy. The pulses are created by sensory transceivers located on or near the body's surface, or embedded in a region of body tissue.

The number of pulses generated by a stimulus in a unit of time follows a particular probability distribution. That distribution is known as the Poisson distribution in honor of the great French mathematician S. D. Poisson who, in 1837, derived the form of the probability distribution from quite primitive assumptions. Later mathematical research showed that the Poisson distribution arises quite frequently in nature because it derives from a much broader set of possibilities than Poisson suspected (Billingsley, 1979; Chow & Teicher, 1988). Most important, the variability in the Poisson distribution is an inherent part of the mechanism that generates pulses of energy. Variability is, part and parcel, a consequence of stimulation, not an error that is added to a pure stimulus.

The transduction of sensory input by Poisson transceivers finds a basis in much earlier work, for example, by Hecht, Shlaer, and Pirenne (1942). But as Crozier (1950) argued, "one does not *see* with the retina" or other peripheral sensory filters, but at an internal site that must be receptive to transduced external stimulation. Comparisons between stimuli appear to be based on a representation of the external stimulus after its reception and analysis at an internal body site, a place where the difference may be "seen."

Peirce and Jastrow (1884) believed that the transmission of peripheral information to an interior body location allowed error to creep into the transmission. This error justified use of the Gaussian Theory of Error in the analysis of sensory discrimination. However, the Poisson mechanism suggests a digital, rather than continuous transmission of information from one to another site within the body. The transmission of a variable number of quanta derived from a Poisson mechanism obviates the need to explain variability as the result of error. Instead, variability is the statistical signature of the transduction mechanism.

Transmission of information requires that the information be transmitted over time. A convenient method of such transmission is the (temporal) Poisson process that emits quantum pulses at a rate that is responsive to a group of spatially arranged transceivers. The average output of the spatially arranged Poisson transceivers controls the average rate of firing of the temporal Poisson process. The concatenation of these spatial and temporal Poisson processes permits the digital transmission of events occurring at the body's surface, or within a volume of tissue, to be translated into a rate of transmission of quantum pulses. At the receiving end, a process of decoding distributes the incoming Poisson-based quantum pulses to receptors that re-encode the primary property (extent and intensity) of the external sensory stimulus.

This digital transmission of information does not require the imposition of Gaussian-based error. Variability is inherent in the transmitted signal as a form of redundancy, not as a matter of error. Nor is there any a priori reason why the transmitted digital signals must be degraded by the addition of error as postulated by Peirce and Jastrow (1884). In spite of Crozier's concerns, there is still good reason to consider the internal representation of the transduced external stimulus to have Poisson properties.

The mathematical properties of the Poisson process, either spatial or temporal, are unusual. Their appeal made the Poisson distribution the basis for many earlier theoretical works by Hecht, Shlaer, and Pirenne (1942), Rashevsky (1938, 1960), McGill (1964, 1967), Creelman (1961), and Laming (1986). A major feature is that the Poisson probability distribution depends only on a single parameter. The single parameter contains all the information about the stimulus and equals the mean number of quantum pulses per unit space or time. An added feature, sometimes the subject of frank criticism, is that the value of the mean equals the variance of the number of quantum pulses.

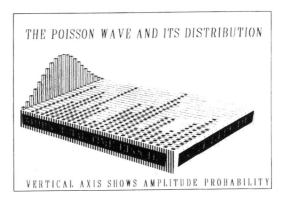

FIG. 5.3. Pulses of energy generated by a stimulus over small units of time, Δt, are collected to provide the amplitude of a Poisson Wave. The ordinate shows the Poisson probability of a particular amplitude. The time axis is defined by equal units of sampling time.

Figure 5.3 illustrates the number of quanta produced by a spatial Poisson process occurring over a unit of space during fixed units of time of size Δt. This three-dimensional image places on the ordinate the probability that a particular number of quanta are emitted during a unit of time, Δt. The number of quanta is discrete so the ordinate actually shows probability. The illustration depicts a mean number of 30 quanta per unit of space occurring during a unit of time. There is only a small probability of as few as 20, or as many as 40, quanta within a unit of time. The ordinate shows clearly the increase and decrease in the Poisson probabilities as the number of emitted quanta increases in value.

Also shown in Fig. 5.3 is the stochastic feature of the process of emitting quantum energy pulses in response to stimulation. The number of emitted quanta is discrete, and the unit of time is constant, but the path from the number emitted during one unit of time to the next creates a *Poisson Wave*. The peaks and valleys of the Poisson Wave correspond to variation in the response to stimulation that is not due to error at all, but is due to the Poisson mechanism that transduces stimulation into quantum pulses.

When two stimuli are to be compared, a Poisson Wave of the type displayed in Fig. 5.3 characterizes each stimulus. The comparison between stimuli, which is postulated to occur at some internal body site, depends on the difference between two such waves, the momentary difference creating a sequence of values that are distributed as the difference between two Poisson variables. Because the distribution of this difference was unknown, previous theories (Creelman, 1961; Laming, 1986; Rashevsky, 1938, 1960) either assumed that the two Poisson distributions were modeled by a Gaussian distribution, or that the difference between two Poisson distributions was essentially Gaussian. These assumptions were

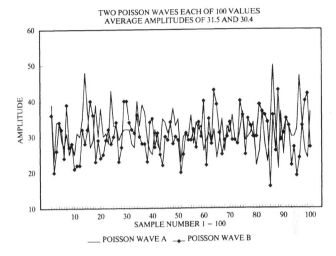

FIG. 5.4. Two Poisson Waves generated by different stimuli do not appear to differ.

needed in order to approximate the form of the probability distribution of the difference between stimuli.

However, approximating Poisson distributions, or their difference, by a Gaussian distribution eases the mathematical intractability at the expense of losing the essential properties of the Poisson mechanism. In fact, the Gaussian approximation obscures just those properties that provide a basis for the frequent occurrence of Weber's Law. Rather than allowing the important information in the difference between two Poisson variables to fall between the boards, so to speak, this new analysis will maintain the difference distribution as a major feature.

Figure 5.4 shows two *Poisson Waves* differing in average value by only a small amount. They correspond to stimulation created by two quite similar stimuli, say S_A and S_B with Poisson Waves W_A and W_B. For each unit of time there is a discrepancy between the waves, but the difference sometimes favors W_A as greater, sometimes W_B. and sometimes neither. By eye there is little chance of determining which wave may be caused by the stronger stimulus. Certainly a single sample will provide little convincing evidence one way or the other.

However, by accumulating evidence, by adding together the successive differences between these waves, an indication of which stimulus is stronger begins to emerge from the variability that clouds perception. As Fig. 5.5 shows, the momentary differences oscillate around zero, indicating little difference between the two waves, as suggested by Fig. 5.4. Yet the accumulated difference grows quickly over time. In short order the accumulation exceeds the response threshold for deciding that Stimulus S_A is the stronger.

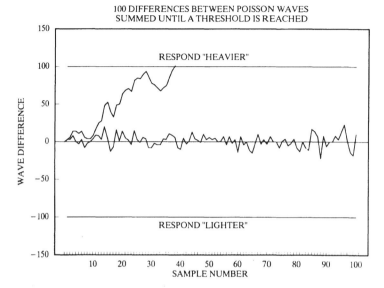

FIG. 5.5. The difference between the two Poisson waves shown in Fig. 5.4 fluctuate around the value zero. The accumulation of these fluctuations quickly overcomes the resistance to respond, A. Unbiased responding is modeled by starting the process of accumulation between the two response thresholds at A and −A.

Predicted Response Proportions

Wave Theory proposes that the accumulation of differences between these Poisson waves is the mechanism that produces decisions about which stimulus is greater, or brighter, or heavier, or in general of larger magnitude than another. Some details of a similar random walk comparison process appear in Link and Heath (1975) and Link (1975), and are summarized by Townsend and Ashby (1983) and Luce (1986) among others. According to the analysis of Link and Heath (1975), response probability depends upon two parameters. First, the amount of resistance to respond, A, is treated as a distance to a threshold value. Second, relative strength of a comparison between two waves is measured by the parameter θ, which is determined by the moment generating function of the Poisson difference wave. The exact details for the existence of the Poisson differences are discussed by Link (1992). Those results show that the unknown parameter θ equals $ln(\mathbf{S_A}/\mathbf{S_B})$ where $\mathbf{S_A}$ and $\mathbf{S_B}$ are the physical values of the stimuli presented for comparison. Notice that this is a dimensionless quantity because the units of measurement for $\mathbf{S_A}$ and $\mathbf{S_B}$ cancel from the numerator and denominator.

For unbiased responding[3], that is, starting the accumulation of wave differences from a point midway between A and $-A$, the probability that the comparison stimulus S_A is judged to be heavier than S_B is

$$Pr\{S_A > S_B\} = \frac{1}{1 + \bar{e}^{\theta A}}. \tag{3}$$

This is the "Logistic" response function often applied by statisticians to biometric and psychological data, but previously lacking a solid theoretical development (Link, 1978). In this instance, the value of θ may be replaced by its equivalent, $ln(S_A/S_B)$, to obtain a general prediction for the probability of choosing S_A to be heavier than S_B.

$$Pr\{S_A > S_B\} = \frac{1}{1 + (S_B/S_A)^A}. \tag{4}$$

Thus, except for the unknown value of the response threshold, A, all response probabilities are predicted to depend upon the ratio of stimulus magnitudes, without any appeal to the Psychological Continuum. The unknown parameter A is somewhat compromising. Perhaps, with a little extra theorizing, even this one remaining parameter can be removed from consideration.

Weber's Law

As early as 1892 Fullerton and Cattell commented that:

> Since Weber published this result, an amount of experiment and theorizing has been devoted to this subject, which has probably never been surpassed in the history of any science. It is commonly claimed that experiments confirm Weber's generalization. At all events, the difference which can just be noticed is usually found to become greater as the stimulus is taken greater. Weber's law can be expressed by the equation
>
> $$N = C \frac{\Delta S}{S}$$
>
> in which N is the least difference which can be noticed, and ΔS the increase in the stimulus S which causes this difference. (p. 21)

Studies of Weber's Law often use the Method of Constant Stimuli, or derivatives, to obtain probabilities of judging comparison stimuli to exceed in magnitude the value of a fixed standard. Regardless of the physical magnitudes of the stimuli presented for comparison, the Gaussian Theory of Error and the Method

---

[3]Link (1991, 1992) also treats the case of response bias and its implications. If biases are symmetric around zero then the Psychometric Function is flattened as if a loss of dicriminability occurred. Naturally a more accurate picture of discriminability develops when the data are corrected for bias, but derivation of Weber's Law presented here remains essentially unchanged.

of Least Squares are used to determine a comparison stimulus value that yields 75% correct responding; what is often called the *threshold stimulus.*

Actually, the estimated threshold stimulus is rarely used as a comparison stimulus. Instead, the Gaussian model is assumed to be true. Then the estimated increment in a comparison stimulus magnitude needed to produce 75% correct responding is plotted as a function of the magnitude of the standard. The result so often reported is that for 75% correct responding the increment in stimulus magnitude versus the magnitude of the standard remains constant. That is,

$$\frac{(S_A - S_B)}{S_B} = \Delta S/S_B \tag{5}$$

$$= \text{Weber's Constant.}$$

This value changes from one sense modality to another and, within a sense modality, from one site of stimulation to another. The Weber Constant for discriminating linear extent is 1/60 but for lifted weights 0.043 (Laming, 1986). The numerical values are generally small, less than 0.10, often less than 0.05, and often showing that about a 3% difference between the standard and comparison stimuli is needed for 75% correct discrimination.

In spite of its frequent measurement in different sense modalities, Weber's Constant remained a mystery for so long it is often forgotten that it must have a source, that it must be a derivable consequence of more fundamental theoretical considerations. Thus it comes as something of a surprise to discover that in those experiments that use a 75% threshold, θ is Weber's Constant.

To derive this famous constant from the quantum transfer of energy by sensory transceivers, notice that the value of θ for the difference between S_A and S_B is, from Equation 2,

$$\theta = ln(S_A/S_B). \tag{6}$$

The difference between these parameters, $S_A - S_B$, equals ΔS, the mean difference between the two stimuli, and also the amount of physical stimulation necessary to increment S_B to S_A. Rewriting θ gives

$$\theta = ln[(S_B + \Delta S)/S_B]$$

$$= ln[1 + \Delta S/S_B]. \tag{7}$$

Expanding in a Taylor's series yields,

$$\theta = [\Delta S/S_B] - (1/2)[\Delta S/S_B]^2 + (1/3!)[\Delta S/S_B]^3 - \ldots.$$

Empirical observation shows that $\Delta S/S_B$ is very much less than one ($\Delta S \ll 1$), often on the order of 0.05, so that the higher order terms [such as 0.5, $(0.05)^2 = 0.00125$] of this expansion contribute little to the numerical value and can be ignored. Therefore,

$$\theta = \Delta S/S_B. \tag{8}$$

This is Weber's Constant.

Weber's Constant determines the amount of added stimulation needed to create a comparison stimulus that can be correctly discriminated on 75% of the experimental trials. Noting that when θA equals 1 the probability of a correct response $Pr\{S_A > S_B\} = 0.731$ suggests that the value of A is the reciprocal of Weber's Constant. When the subject discriminates correctly with probability 0.731, close enough to 0.750 for practical purposes[4], then

$$\theta A = 1$$

or,

$$A = \frac{1}{\theta} . \tag{9}$$

The unknown value for θ is Weber's Constant as determined by studies on the discriminability of lifted weights. Laming (1986), on the basis of a reanalysis of Oberlin's (1936) study of Weber's Law for weight discrimination, suggests the value $\theta = 0.043$. From this A equals 25.6. Using $A = 25.6$ to predict the response proportions in Guilford's experiment gives a completely constrained prediction of response proportions. As shown in Fig. 5.6, the predicted and observed proportions agree quite well. Seven comparison stimuli are compared against seven standards. For each standard the predicted probabilities of responding "Heavier" are connected by a solid line. The observed response proportions are shown as squares. The correspondence between theory and data is quite close. The chi-squared value for goodness of fit equals 31.0, based on 28 degrees of freedom, and no estimated parameters. Not only has the Psychological Continuum vanished, but with it goes the need to estimate unknown parameters.

Decision Time Predictions

For technical reasons, response times for judgments of lifted weights are difficult to measure. Typically the hands are used to lift the weights, thereby precluding their use in pressing choice keys that may stop clocks used to measure choice response time. Guilford's failure to measure response times is understandable. Yet, decision times did occur, and although they are unreported we can use the predictions of Wave Theory to determine the unobserved response times.

Wave Theory predicts that, for unbiased responding, the average time taken to

---

[4]Actually the equation for Weber's Constant based on the Gaussian model is often written in the form $\Delta S/(S_B + C)$ where C is a constant. In other words the value should be smaller than $\Delta S/S_B$. This more refined form is actually in keeping with the idea that a just noticeable difference occurs when the correct response probability equals 0.73 rather than 0.75. Link (1992) provides additional details.

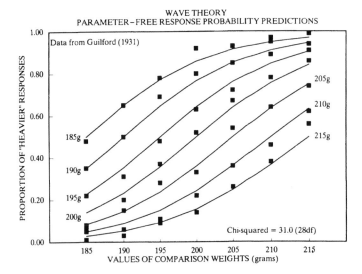

WAVE THEORY
PARAMETER–FREE RESPONSE PROBABILITY PREDICTIONS

FIG. 5.6. Parameter-free predicted response proportions agree closely with the observed proportion of "heavier" judgments found by
Guilford (1931).

decide whether the Comparison Stimulus S_A is heavier or lighter than the Standard S_B is,

$$DT_{AB} = \frac{A\ Pr\{S_A > S_B\} - A\ Pr\{S_B > S_A\}}{k(S_A - S_B)} \qquad \text{when } S_A - S_B \neq 0, \qquad (10a)$$

$$= \frac{A^2}{k(S_A + S_B)} \qquad \text{when } S_A - S_B = 0, \qquad (10b)$$

where k is a constant of proportionality.

The values of response probabilities, the resistance to respond, A, and the magnitudes of the stimuli are all known. Thus the predicted, but unmeasured, decision
times can be calculated directly. The actual times are relative to a unit of time for
accumulating differences between Poisson waves. Although this rate is unknown, these relative decision times reveal correctly the relative changes in
decision time.

Figure 5.7 shows the decision time surface. The decision times decline, as is
to be expected, when the difference between S_A and S_B increases. The relative
units of time used to scale the ordinate show a quite dramatic decline as the
comparison moves away from the main diagonal. Although the surface appears
to the eye to be symmetric, slight differences exist for symmetric off-diagonal
elements due to a corresponding asymmetry in the response proportions. Never-

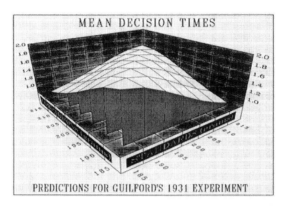

FIG. 5.7. Predicted decision times (in relative units of time) for Guilford's judgments of lifted weights. The predictions are entirely parameter free.

theless, the results are quite in keeping with response times obtained from other psychophysical studies of discrimination (Link, 1992; Luce, 1986).

A surprise is hiding in these predicted decision times. The mean decision time when no difference is presented, when S_A equals S_B, are the largest for a fixed Standard, as suspected long ago by Cattell (1902). As the comparison deviates farther from the standard, the decision times decrease, as shown by the off-diagonal elements. More significant is the result of increasing stimulus magnitude. Generally, as the magnitudes of S_A and S_B increase, the predicted decision time decreases. When there is no difference between S_A and S_B, the decision times also decrease. Although no difference exists between S_A and S_B, the variance of the corresponding waves, W_A and W_B becomes larger. Increased variability leads to larger deviations from zero and the accumulation of larger deviations. Although the response proportions will show no change, remaining at 0.50 because there is no response bias and the stimuli do not differ, the decision times become faster as stimulus magnitudes increase. This tantalizing prediction is now searching for an appropriate proving ground.

SUMMARY

In keeping with the foundations of Stimulus Sampling Theory (Estes, 1950), Wave Theory postulates that sensory transceivers project an image of the external world to internal body sites. A new theory of this transmission, based upon spatial and temporal Poisson processes, shows how a single Poisson parameter captures essential information about events at the peripheral location. When stimuli are compared against each other, each stimulus evokes a characteristic wave determined by the Poisson parameter. These waves are compared by subtraction, through the use of excitatory and inhibitory inputs to a wave accumulator. The accumulation proceeds until either an excitatory or inhibitory resistance to respond (a response threshold) is first reached. The threshold reached first determines which response the subject chooses.

Wave Theory makes no appeal to error as an explanation for the failure to discriminate. Rather, discrete receptive units, transceivers, combine their responses to sensory stimulation to produce electrical signals. Differences between these signals drive the subject's decision process. Mathematical analysis of the decision process shows that the response probabilities produce Logistic response functions that are determined by the subject's resistance to respond, A, and the measure of discriminability θ. The relation between θ and stimulus magnitude yields a general psychophysical law that includes Weber's Law as a special case. Because the value of θ also proves to equal $\Delta S/S$, the value of θ is Weber's Constant. But Weber's Constant was shown to equal the reciprocal of the subject's resistance to respond, that is, $\theta = 1/A$. In this way Wave Theory provides a new foundation for, and interpretation of, Weber's Law and Weber's Constant.

Guilford's (1931) paired comparison study of lifted weights provides a convenient testing ground for parameter-free predictions of response proportions. Without estimating any parameters from Guilford's data the chi-squared statistic for goodness of fit proves to equal 31.0 based on 28 degrees of freedom—a remarkably good set of predictions given that no parameters are estimated from the data. Although Guilford did not measure response times, Wave Theory predicts the decision times that Guilford did not observe. These parameter-free predictions agree well with the general findings for discriminative judgments, and suggest new features that await future testing at another proving ground.

REFERENCES

Billingsley, P. (1979). *Probability and measure*. New York: John Wiley & Sons.

Cattell, J. M. (1902). The time of perception as a measure of differences in intensity. *Philosophische Studien, 19*, 63–68.

Chow, Y. S., & Teicher, H. (1988). *Probability theory: Independence, interchangeability, martingales* (2nd ed.). New York: Springer-Verlag.

Creelman, C. D. (1961). Human discrimination of auditory duration. *Journal of the Acoustical Society of America, 34*, 582–593.

Crozier, W. J. (1950). On the visibility of radiation at the human fovea. *Journal of General Physiology, 34*, 87–136.

Estes, W. K. (1950). Toward a statistical theory of learning. *Psychological Review, 57*, 94–107.

Fechner, G. T. (1966). *Elements of psychophysics* (H. E. Adler, Trans.) New York: Holt, Rinehart, & Winston. (Original work published 1860)

Fullerton, G. S., & Cattell, J. M. (1892). *On the perception of small differences*. Publications of the University of Pennsylvania. Philosophical Series. No. 2. Philadelphia: University of Pennsylvania Press.

Gauss, C. F. (1809). *Theoria motus corporum coelestium*. Hamburg: Perthes and Besser.

Gauss, C. F. (1880). Theoria combinationis observationum erroribus minimis obnoxiae, pars prior. In *Carl Friedrich Gauss Werke. Vol. 4* (pp. 1–26). Göttingen: Königlichen Gesellschaft der Wissenschaften zu Göttingen. (Original work published 1821)

Guilford, J. P. (1928). The method of paired comparisons as a psychometric method. *Psychological Review, 35,* 494–506.

Guilford, J. P. (1931). Some empirical tests of the method of paired comparisons. *Journal of General Psychology, 5,* 564–577.

Hecht, S., Shlaer, S., & Pirenne, M. H. (1942). Energy, quanta, and vision. *Journal of General Physiology, 25,* 819–840.

Laming, D. (1986). *Sensory analysis.* New York: Academic Press.

Link, S. W. (1975). The relative judgment theory of two choice response time. *Journal of Mathematical Psychology, 12,* 114–135.

Link, S. W. (1978). The relative judgment theory of the psychometric function. In J. Requin (Ed.), *Attention & performance VII* (pp. 619–630). Hillsdale, NJ: Lawrence Erlbaum Associates.

Link, S. W. (1990). A waveform theory foundation for Weber's, Fechner's, and Stevens' laws. Abstracts of the 1990 Annual Meeting of the American Association for the Advancement of Science, 49–50.

Link, S. W. (1992). *The wave theory of difference and similarity.* Hillsdale, NJ: Lawrence Erlbaum Associates.

Link, S. W., & Heath, R. H. (1975). A sequential theory of psychological judgment. *Psychometrika, 40,* 77–105.

Luce, R.D. (1986). *Response times.* New York: Oxford University Press.

McGill, W. J. (1964). *Introduction to counter theory in psychophysics.* Unpublished manuscript. Stanford University, Institute for Mathematical Studies in the Social Sciences, Stanford CA.

McGill, W. J. (1967). Neural counting mechanisms and energy detection in audition. *Journal of Mathematical Psychology, 4,* 351–376.

Neimark, E. D., & Estes, W. K. (1967). *Stimulus sampling theory.* San Francisco: Holden-Day.

Oberlin, K. W. (1936). Variation in intensitive sensitivity to lifted weights. *Journal of Experimental Psychology, 19,* 438–455.

Peirce, C. S., & Jastrow, J. (1884). On small differences in sensation. *Memoirs of the National Academy of Sciences, 3,* 73–83.

Poisson, S. D. (1837). *Recherches sur la probabilité des jugements en matière criminelle et en matière civile, précedées des règles générales du calcul des probabilitiés.* Paris: Bachelier.

Rashevsky, N. (1938). *Mathematical biophysics.* Chicago: University of Chicago Press.

Rashevsky, N. (1960). *Mathematical biophysics: Physico-mathematical foundations of biology* (3rd ed.), Vol. 2. New York: Dover Publications.

Thurstone, L. L. (1927a). Psychophysical analysis. *American Journal of Psychology, 38,* 368–389.

Thurstone, L. L. (1927b). A law of comparative judgment. *Psychological Review, 34,* 273–286.

Townsend, J. T., & Ashby, F. G. (1983). *Stochastic modeling of elementary psychological processes.* New York: Cambridge University Press.

6 A Mathematical Theory of Attention in a Distractor Task

David LaBerge
University of California, Irvine

The notion of attention as the operation by which a part of the total stimulus information comes to receive specialized processing bears strong resemblance to the stimulus sampling notion first set forth by Estes (1950) in his groundbreaking paper on statistical learning theory. At that time, the conceptual foundation of mainstream psychology resided in the area of learning, and the prevailing learning theories assumed that the organism unselectively takes in all information provided by the momentary stimulus. Estes departed from the assumption of processing the "whole stimulus" when he assumed instead that only a "sample" of the available information is processed. The response connections of the sampled stimulus elements led to the generation of a response probability, and the reconnecting of the sampled stimulus elements to one of the response classes represented the learning process. The elegant mathematical formalization of these processes, built on the disarmingly simple assumption of stimulus sampling, gave rise to his well-known quantitative predictions of the course of learning in choice situations (e.g., Estes & Straughan, 1954; Neimark & Estes, 1967).

The major difference between the traditional stimulus sampling notion and the selective attention notion lies in the matter of control. For purposes of deriving learning curves one could allow the sampling across a population of stimulus elements to be completely random, with the consequence that the ensuing mathematical derivations could proceed with few parameters in a relatively straightforward manner. However, in the case of selective attention, it is typically assumed that the sampling of stimuli is nonrandom, that is, the sampling probability is biased toward a particular part of the stimulus array. The sampling bias is assumed to be controlled from some source external or internal to the organism.

External sources of control (sometimes called "bottom-up" sources, "involuntary attention," "pop-out," or "attentional capture") induce the organism to process particular spatial locations that have greater luminance, or a different color or movement velocity as compared to the surrounding locations. Internal sources of control (sometimes called "top-down" or "voluntary attention") direct the organism to process particular locations, colors, orientation features, or velocities, according to cognitive procedures or actions based on higher-order processes. Thus, for attention, external and internal factors exert control over the part of the momentary stimulus input selected for some kind of specialized processing.

Expressed more formally, completely random stimulus sampling operates on a uniform probability density function defined over the space of available stimulus elements, whereas attentional selection typically operates on a nonuniform probability density function defined over the space of available stimulus elements. Presumably, the most frequent type of density function represented by attentional processing is the single-peaked distribution, which has sometimes been labeled an "attention gradient." However, mathematically speaking, the term "gradient" denotes "slope," and as a slope does not indicate a modal value, it has recently been suggested (LaBerge, 1990a, 1990b) that the term "peaked activity distribution" be substituted for the term "gradient."

The single-peaked activity distribution of attention is assumed to represent the rate of information flow (or activity) at various locations across a space, where the space is defined in terms of spatial coordinates or in terms of similarity distances, (see Shepard, 1989) such as exist for shapes or colors. The peak of the distribution is located at the point on a continuum representing the expected stimulus attribute or location. In the case of attending to spatial location, a target location is selected by higher-order processes, and subsequently activity at the selected point is increased by means of reciprocal flow from a central filtering mechanism (LaBerge, 1990b). As the enhanced information flow builds activity at the target point, the activity spreads in the neighborhood of the expected point, thus producing a unimodal distribution, somewhat like erecting a tent by pushing upward at the center of the tent fabric with a tent pole.

The main purpose of the present paper is to describe a model of the peaked activity distribution that represents a concentration of attention at a narrow target area, such as concentrating one's attention at the location of a particular letter within a word. The model of attention is applied to the experimental task of identifying a letter in the presence of distracting letters. It turns out that this type of "target-plus-flankers" task provides a relatively direct measure of attention spread by means of response time measures (Eriksen & Eriksen, 1974; LaBerge, et al., 1991). However, to generate quantitative predictions of response time data, it is helpful to interpose a response time model between the attention-spread model and the response measures.

The plan of the paper, then, is to (a) describe the target-flanker task and the measure of attention spread that it provides; (b) summarize two sets of data that apparently indicate how the spread of attention may be narrowed experimentally; (c) propose a mathematical model of the peaked activity distribution representing the spread of attention; (d) describe how the selected spatial location at any moment gates a particular letter in the display and allows a set of stimulus elements to enter the shape identifier; and (e) derive predictions for the present data based on the attention model combined with each of two models of response generation.

MEASUREMENT OF ATTENTION SPREAD BY THE TARGET-FLANKER TASK

One way to obtain a behavioral indicator of attention spread is first to induce the subject to concentrate attention over a specific area of the visual field corresponding to the size and location of an expected target, and then to determine whether or not stimuli located in the surrounding areas are processed. If items in the surround show no effects on the response measure, then one is led to conclude that the system did not process information in the surround, and therefore that attention spread was constrained to an area equal to or less than the area of the target stimulus. Of course, control conditions must show that the stimulus items in the surround have the potential to affect the response measure if they were processed.

The present measure of attention spread is based on the procedure developed by Eriksen & Eriksen (1974), and bears close similarity to procedures of Estes (1972) and Kahneman and Chajczyk (1983). Four letters, C, H, S, and K were assigned to two responses, two letters to a response, as shown in Fig. 6.1.

DISPLAY CONDITIONS FOR TARGET-FLANKER TASK

RESPONSE ASSIGNMENTS		DISPLAY TYPES	DISPLAYS
C	S	Incompatible	SSSSSSSSHSSSSSSSS
H	K	Compatible	CCCCCCCCHCCCCCCCC
left button	right button	Same	HHHHHHHHHHHHHHHHH
		Neutral	XXXXXXXXHXXXXXXXX

The measure of attention spread is RT(incompatible) - RT(compatible),
abbreviated I - C.
(If the width of spread is one space, then I - C = 0) .

FIG. 6.1. Four types of stimulus displays presented in the target-flanker task. Subjects were instructed to identify the center letter of a display by pressing one of two buttons. An example of one of the pairing of letters with a response is shown in the left of this figure.

Stimulus displays were horizontal strings of 17 characters, with the target letter always at the center location. The flanking letters that were placed in the horizontal surround of the target letter were of four types: response incompatible, neutral, response compatible, the same as the target letter. The measure of attention spread was the response time difference between the Incompatible and Compatible flanker displays (which we call "I-C"). Eriksen and Eriksen varied the separation between letters within a display and found that I-C varied from approximately 80 ms at the closest spacing (0.06 degree) to 20 ms at the farthest spacing (1.0 degree). A positive value of I-C indicates that subjects had identified the flanking letters along with identifying the target letter, so that when between-letter spacing was 1.0 degree, subjects apparently had spread attention over a distance of at least 1.0 degrees plus two letter widths. Presumably, if spacings were increased sufficiently, subjects would not have processed the flanking letters at all, and the I-C value obtained would have been zero.

Attempts to Control Attention Spread

With this measure of attention spread in hand, we looked for ways to narrow the spread of attention while maintaining a constant between-letter spacing (0.14 degree). Our approach was to vary the characteristics of a target item that was presented just prior to the flankered display, as illustrated in Fig. 6.2 (LaBerge, 1983). A trial contained two targets, in immediate succession, and the subject was instructed to identify both targets before pressing the button. Two variations in presentation of the first target display appear to affect I-C values: changing the duration of the first target, and changing the relative size of the first target. When the duration or size of the first target is decreased, the subject presumably reduces the spread of attention around the area of the target to maximize sampling of target features in the limited time in which the target is displayed.

PROCEDURE TO CONTROL THE SPREAD OF ATTENTION

	TRIAL EVENTS	DURATIONS (ms)
WARNING SIGNAL	# # # # # # # # * # # # # # # # #	1000
TARGET 1	ZTZTZTZT7ZTZTZTZT	67, 100, 150, 250, 350, 450
TARGET 2	SSSSSSSSHSSSSSSSS	200
MASK	# # # # # # # # * # # # # # # # #	300
CATCH DISPLAYS FOR TARGET 1	ZTZTZTZTTZTZTZTZT	
	ZTZTZTZTZZTZTZTZT	
	TZTZTZTZZTZTZTZTZ	
	TZTZTZTZTTZTZTZTZ	
RESPONSE RULE :	Respond to the center letter in Target 2 only if a 7 occurs there first.	

FIG. 6.2. Examples of the four displays of a trial. The displays were presented successively and in the same location. Target 2 was one of the four display types of Fig. 6.1, and the display types were randomized within a block of trials.

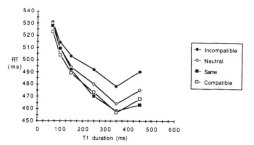

FIG. 6.3. Mean response time to the stimulus displays of Fig. 6.1 shown to subjects according to the procedure depicted in Fig. 6.2. Data from LaBerge et al. (1991).

Reducing attention spread lowers the sampling of flanker features in the subsequent second target and thereby decreases the value of I-C.

The data from an experiment that presented 6 different durations of the first target are shown in Fig. 6.3 (LaBerge, Brown, Carter, Hartley, & Bash, 1991). Mean response times of 24 subjects to the Incompatible and Compatible displays show a significant convergence. Errors and misses were below 1.5% except at the Target 1 duration of 67 msec, where errors increased to approximately 13% and misses increased to approximately 4.5%. This increase in errors and misses precluded using durations less than 67 msec in an effort to reduce I-C values to less than 9 msec.

We also attempted to narrow attention spread by reducing the size of Target 1 relative to the size of Target 2. In one condition the letters in Target 1 displays were constructed to fill a 3 by 5 matrix of pixels, and in the other condition the letters in Target 1 were constructed to fill the usual 5 by 7 matrix. The characters of Target 2 and the warning signal were constructed to fill a 5 by 7 matrix. The results from the 46 subjects of this unpublished study are shown in Table 6.1. The convergence of response times to Incompatible and Compatible displays was significant statistically, and was interpreted as indicating a reduction in the spread of attention. It is this convergence in I-C that we wish to derive from the mathematical model to be described later.

TABLE 6.1
Mean Response Times to Incompatible and Compatible Flanker Displays As a Function of Size of First Target Letters

	Type of Display		
	Incompatible	Compatible	Difference
Size of Letter			
Large	616	593	23
Small	647	633	14

Another salient feature of the present data, aside from the convergence of I-C, is the increase in response time as duration and size of the first target is reduced and I-C decreases. Apparently, the obtained decrease in I-C is not simply due to the increase in absolute response times because an additional experiment that varied response time by a deadline manipulation did not show a decrease in I-C as response time increased (LaBerge et al., 1991). A possible explanation for the overall increase in response time is that the rapid sequence of two target items induces a degradation in stimulus information from each, hence increasing the time required to identify them. Another explanation is that the rapid sequencing of the two targets produces an intrusion of the processing time of the first target into the display time of the second target, and therefore the actual initiation of processing of the second target is delayed beyond its objective onset as the duration of the first target decreases. In any case, the obtained change in absolute response time is assumed in the present model to be independent of change in the spread of attention.

Other notable aspects of the data shown in Fig. 6.3 include the finding that the Compatible and Same displays produce quite similar response times except possibly at very short durations of Target 1. The higher similarity between target and flankers in the Same displays could produce a lateral inhibition effect (see Bjork & Murray, 1977) that increased at short exposures, leading to a longer processing time for Same over Compatible displays in this region of the curves.

Of particular relevance to the response time models described in this paper are the response times evoked by the Neutral displays relative to those evoked by the Incompatible and Compatible displays. It is clear in Fig. 6.3 that the Neutral curve lies between the curves of the Incompatible and Compatible displays, and perhaps closer to the Compatible curve. The positioning of the Neutral display curve may depend upon how similar the Neutral letters are to the Compatible and Incompatible letters. The particular Neutral letter used in this experiment was the letter X, which shares features with the K, and half of the time the K served as a Compatible letter and the other half of the time it served as an Incompatible letter. However, in a subsequent study, we used X as a Neutral letter against response-evoking letters of C, H, D, and F, in order to render the X more neutral. Using T_1 and T_2 durations of 200 msec, we obtained mean response times of 563 for the Incompatible display, 542 for the Compatible display, and 553 for the Neutral display. Again, the response time for the Neutral display appears to lie somewhere midway between the response times to the Incompatible and Compatible displays.

We turn now to a theoretical treatment of two major processes involved in the present shape identification task: the variation of attention spread, and the generation of response times. These two processes will be connected by means of a model of the stimulus elements representing information in a stimulus display. The attention model describes the elements in a stimulus display that are sampled moment-by-moment, and the response time models describe alternative views of how the sampled elements could be converted into response frequencies and response times. The model of the stimulus elements is in all major respects the type of model developed by Estes in his 1950 paper, and employed by him and others in a great many related papers since then (e.g., Estes & Burke, 1953; LaBerge, 1959; for others, see Neimark & Estes, 1967).

Theory of Attention Operations in Shape Identification

The I-C data in Fig. 6.3 and Table 6.1 indicate that during a typical exposure of a display the flanker items are occasionally identified along with the target item. Assuming that the extent of attention spread determines how often flanker items are identified during a display, there are two principal ways to represent attention spread. The first way is to assume that the size of attention spread varies, but its location, or central tendency, remains fixed. All items falling within that area are simultaneously identified, with strengths varying according to the corresponding ordinate value of the activity distribution. This case can be called a *parallel attention model*. The second way to represent attention spread, adopted here, is to assume that the size of attention is tailored to the size of a single item, but the location of attention rapidly fluctuates across target and adjacent flanker locations. This *serial attention model* is based on the theory of attention in shape identification described by LaBerge & Brown (1989) that assumes that, for the present type of task, subjects identify only one shape at a time (see Fig. 6.4). Therefore, when the spread of attention increases or decreases, it is implied that the range of fluctuation of a relatively small attention area (corresponding to the size of one character) increases or decreases. More precisely, the distribution of locations of the channel location expands or contracts around a central peak located at the target as the spread of attention changes.

It will be assumed, therefore, that a rapid fluctuation of the position of the filter channel can be represented by probabilistic sampling of positions at and

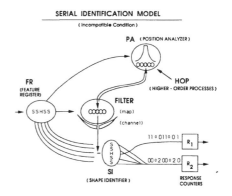

FIG. 6.4. Schematic representation of the cognitive domains involved in identifying a familiar shape in a cluttered field, and a representation of the response counters that link an identification event to both response choice and response time. The large O's entering the response counters are neutral elements arising from the stimulus display; the small O's are background (noise) elements.

SERIAL IDENTIFICATION MODEL
(Incompatible Condition)

PA (POSITION ANALYZER)

HOP (HIGHER - ORDER PROCESSES)

FR (FEATURE REGISTER)

FILTER (map)

SSHSS

(channel)

11 o 0 11 o 0 1   R₁

00 o 2 00 o 2 0   R₂

SSHSS

SI (SHAPE IDENTIFIER)

RESPONSE COUNTERS

around the target position. The probabilistic sampling is determined by a probability density function defined over a map of positions corresponding to the item positions in the stimulus displays shown in Fig. 6.1. This probability density function is assumed to be located in the Position-Analyzer Domain (see Fig. 6.4). Task instructions, operating through higher-order-processes (HOP) select a given position in the displayed string of items (usually the center position, suitably indicated in the warning signal) represented in the PA domain. Subsequently, activity at this location quickly builds up, owing to an enhancement circuit that connects this PA location with a corresponding location in the filter (LaBerge, 1990). But activity at the central location of the PA map spreads to surrounding regions, so that there exists some uncertainty that the PA domain will always select the target's location over a flanker's location. Thus, the area under the position density function corresponding to each letter's position in the display represents the probability that a particular letter position in PA will direct information flow to the corresponding location in the filter map. An enhancement of activity in the filter map will, in turn, augment information flow in the corresponding location of the feature information entering the shape identification domain (see Fig. 6.4).

Model for the Spread of Attention

The spread of attention in the neighborhood of the target letter in the flanker task is assumed to be represented by a spatial distribution of activity or resources. When the subject anticipates that the target item will appear in a given location, higher-order processes, assisted by reciprocal connections between the PA domain and the filter, act to increase the rate of information flow at this target location. The rate of flow at the target's location in the PA domain then diffuses or spreads to neighboring locations, such that the rate of flow at neighboring locations decays as a function of distance, x, from the target's location, according to the differential equation

$$\frac{dH(x)}{dx} = -aH(x), \qquad \text{where } 0 \leq x \leq \infty, \text{ and} \qquad (1)$$

$H(x)$ is the rate of information flow at location x, and a is a positive constant.

For the present situation in which flankers appear only to the right or left of the target, we assume that the decay rate is symmetrical, so that $x = |x|$. The solution to this equation is

$$H(x) = H(0) \exp(-ax). \qquad (2)$$

For simplicity of notation we let

$$h = H(0), \text{ so that}$$

$$H(x) = h\exp(-ax). \qquad (3)$$

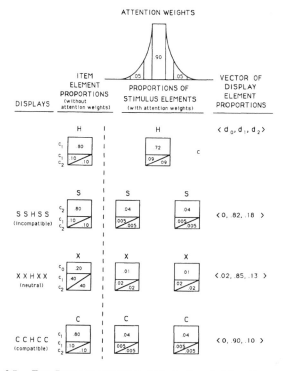

FIG. 6.5. *Top.* Example of back-to-back exponential functions representing the decay portion of attention-spread over the locations of items in a display. The numbers within a sector beneath the curve denote the probability that this sector is sampled in a given time unit. Probability values for a location more than one unit removed from center are negligible under present parameter settings. *Bottom:* Examples of stimulus element proportions for each item in a display. The center item is shown in the first row of boxes, and the flanker items are shown in the lower three rows of boxes. The similarity (overlap of elements) between items is assumed to be .20. When the element proportions of each item are combined into a string of items, the summary proportion is denoted by a three-element vector, shown in the right column of the figure. The components, d_1, d_2 and d_0, denote element proportions connected to Response 1, Response 2, and neither response, respectively.

An illustrative example of this back-to-back exponential distribution of activity across a stimulus display is shown at the top of Fig. 6.5. The greater the "concentration of attention" given to the center location, the higher the activity at that location, and the steeper the slope of the curve on either side of that center location. Assuming that the parameter h varies systematically with an independent variable such as the duration or size of the first target in the present experimental situation, we employ the subscript i to represent the value of the independent experimental variable. Therefore, the more general form of Equation (3) becomes

$$H_i(x) = h_i \exp(-a_i x). \tag{4}$$

We let b denote the width of the representation of a target item so that the area, A, under the curve of Fig. 6.5 corresponding to the total activity at the target location, t, is a rectangle with area

$$A_{ti} = b_i h_i. \tag{5}$$

We shall assume that the probability of sampling the target location at a given time can be expressed as the ratio of the center rectangular area to the total area under the curve. In effect, we convert the distribution of activity into a probability density function. Thus the probability that a particular item position will send a line of flow to the corresponding position in the filter map is represented by the area under the density function above the location of the particular item.

To derive an expression for the probability density function from Equation (4), we need to express the area above the flanking letters. Fortunately, because the flanking letters in a given display in the present experiment are identical, we do not need to express areas for each of the several flanking positions separately, but can proceed directly to the sum of all the areas above flanking letters. We denote the letter widths (5 pixels) as b, and the separation between letters (3 pixels) as s. The total width of a display of 17 letters is therefore $17b + 16s$ (133 pixels). Let A_{fij} denote the area above the jth flanker position for the ith experimental condition; then the sum of all of the flanker areas on one side is

$$\sum_j^n A_{fij} = \int_s^{s+b} h_i \exp(-a_i x) dx + \int_{2s+b}^{2(s+b)} h_i \exp(-a_i x)\, dx + \dots$$

$$\int_{n(s+b)-b}^{n(s+b)} h_i \exp(-a_i x) dx = (2h_i/a_i) S_i(n), \tag{6}$$

where

$$s_i(n) = \frac{\{\exp(-sa_i) - \exp[-(s + b)]\}\{1 - \exp[-n(s + b)a_i]\}}{1 - \exp[-(s + b)a_i]} \tag{7}$$

For the present experimental displays of eight flankers on a side, the quantity

$$\exp[-n(s + b)a_i] = \exp[-8(s + b)a_i]$$

is negligible (for appropriate values of a), so that the last term in the numerator of Equation (7) can be treated as unity.

We now use the expressions we have obtained for the center rectangular area

(Equation (5)) and the entire flanker area on one side (Equation (6)) to obtain the probability that the central target location will be sampled (i.e., augmented) by the filter, and the probability that one of the flanker locations will be sampled by the filter. For the target location, the sampling probability is

$$q_{ti} = \frac{A_{ti}}{A_{ti} + 2\sum_j^n A_{fij}} = \frac{bh_i}{bh_i + 2h_iS_i(n)/a_i} = \frac{ba_i}{ba_i + 2S_i(n)}, \tag{8}$$

and for any of the flanker locations, the sampling probability is

$$q_{f_i} = 1 - q_{ti}. \tag{9}$$

We note that the center height parameter, h_i drops out of this expression. However, one may preserve h_i in the equation by adding the assumption that the area under the activity curve is constant for all values, i, of the independent variable. That is,

$$A_{ti} + 2\sum_j^n A_{fij} = K, \qquad \text{for all } i. \tag{10}$$

Since

$$A_{ti} = bh_i \text{ and}$$

$$2\sum_j^n A_{fij} = 2b_iS_i(n)/a_i,$$

then

$$K = bh_i + 2h_iS_i(n)/a_i$$

and

$$h_i = \frac{K}{ba_i + 2S_i(n)} \tag{11}$$

Thus, the expression for q_{ti} can be used when we wish to speak of the degree of concentration of attention at the center item, and the expression for q_{f_i} can be used when we wish to speak of the degree to which attention is spread beyond the center item. Of course, one term is simply the complement of the other, because at any moment during the registration of the display, it is assumed that the filter channel is aligned either at the center location or at one of the flanker locations.

Model of Stimulus Elements

We now turn to a specification of the content of the stream of informational elements flowing to the response counters, as illustrated in Fig. 6.4. Following the stimulus sampling notions of Estes (1950), it is assumed that the content of the set of elements entering the response counters is based on the learned connection between a stimulus item and a particular response counter. Therefore, our representation of the content of each letter in a display is described in terms of the response elements generated by that letter. In addition, we must take into consideration the likelihood that the letter items used in typical experiments will bear some degree of similarity to each other. We interpret similarity to imply that, on occasion, the presentation of one letter item could result in the identification event corresponding to another letter item. For example, the presentation of the letter H could be mistakenly identified as a K or, with lower probability, a C. Similarly, the letter X could occasionally be identified as a K. We specify the degree of similarity of two items by assuming that the information arising from each letter display contains two disjoint sets of elements: a set that is unique to that letter, and a set that is common to the other letters used in the task. The unique set contains response counter elements of the type assigned to that letter by instructions and learned to a high level by practice. The common set contains response elements divided equally between the two response types.

The identification of an item depends upon the content of the feature information in the selected sample. The attention probability density determines which part of the total display information is sampled. The features of the total display are first registered in the Feature Registration Domain (see Fig. 6.4) and information then flows toward the Shape Identification Domain. But before this typically rich array of information reaches the Shape Identifier, it is filtered by means of the intersection of a topographically constrained flow from the Filter Domain. The location within the display at which the connection from the filter enhances the flow of information into the Shape Identifier depends probabilistically on the peaked density function that has just been described.

We now relate the density model of attention spread to the content of the stream of elements entering the response counter elements by specifying the elements that arise from each letter location in a display. For example, in the case of an Incompatible display, the target item in the center of the display presumably gives rise mainly to elements that are connected to one response, and the flanker items produce a majority of elements that are connected to the other response. For the Compatible and Same flanker displays, target and flanker items produce elements connected mainly to the same response. For the Neutral flanker display, the target item is assumed to produce a set of features that are predominantly of one response type, but the neutral flankers are assumed to produce some proportion of features that are connected to no response, with the remaining proportion divided between the two responses.

Representation of a Display Item as a Set of Stimulus Elements

An illustrative example of how the set of elements of target and flanker items might be partitioned is given in the left column of boxes in Fig. 6.5. In each box,

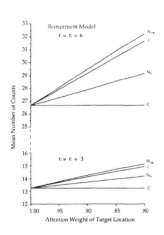

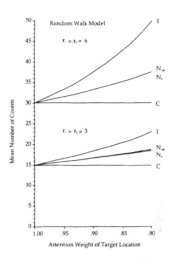

FIGS. 6.6. and 6.7. Mean numbers of elements sampled serially by the two response counters until a counter first records a criterion count (r) of elements of that response class, as predicted by the Independent Counter (Recruitment) Model and the Interactive Counter (Random Walk) Model.

the upper rectangle represents the set of elements unique to that letter, and the lower rectangle represents the set of elements held in common with the other letters. For the target item, the unique elements are assumed to be connected to response type 1, and the set proportion is labeled c_1. The common element set is divided equally across the two response types, and so half of the set contains c_1 elements and half c_2 elements. For a flanker item of the Incompatible display, the elements in the unique set are connected to the other response and so are labeled c_2. For a neutral flanker item, the elements of the unique set are connected to neither response, and are labeled c_0 (neutral elements).

Element-Set Partitioning for a Display as a Whole

Given the element content of the individual letter items, we can derive the element content of a given display by averaging the proportions of element types across one target plus 16 flankers. But if in the averaging operation we weight each of the 17 letter items equally, we would obtain a set of counter elements that would lead to wrong, even grotesque predictions of both response accuracy and response time. For example, in an Incompatible flanker display, the vast majority of elements are connected to the response opposite to that of the target; hence performance should approach 100% errors. Furthermore, in a Neutral flanker display, the overwhelming number of neutral elements arising from the flankers would produce mean response times far greater than those for the Incompatible and Compatible displays. Therefore, to give an appropriate account of the data from these displays, we must assume some weighting function defined on the locations of the items such that the target item receives relatively high weight, and the flanker items relatively low weights. Attention provides this function.

For the present theoretical description, the variation in sensitivity across retinal locations will not be a parameter in the weighting equations, mainly because it turns out that the weights to be assigned to flankers beyond the two adjacent to the target item are near zero, so that the central three items contribute virtually all the stimulus information in these displays, as illustrated at the top of Fig. 6.5. These three items together subtend a little over a degree of arc, which is an area covered for the most part by the fovea.

Applying the Attentional Weighting Function to Display Element Sets

The specific weighting function that we apply to the individual letter items of the display has already been derived in a previous section. Its mirrored-exponential form is shown at the top of Fig. 6.5, where, for purposes of illustration, the weight for the target location is assumed to be .90 and the weights for the adjacent flanker items are each .05. These proportions correspond to a particular display size or a particular duration of Target 1 (the ith experimental condition). When these weights are multiplied with the element proportions given in the first column of boxes in Fig. 6.5, the contents of the item boxes are transformed into the box contents given in the three central columns of Fig. 6.5. The three central columns (under "Proportions of Stimulus Elements") represent, from left to right, the flanker to the immediate left of the target, the target, and the flanker to the immediate right of the target (the target is the same for each of the flanker types, here, so only one box appears in the middle column). In order to obtain the element content (partitioning) of a given display that enters the response counters, we average the proportions across these three boxes. The resulting proportions for the three types of elements for each flanker type are given in the last column of Fig. 6.5. Thus, the informational content of a display delivered to the response counters after it has been weighted by spatial attention is expressed by the vector $<d_0, d_1, d_2>$, where the d_k, $k = 0,1,2$, represent the proportion of elements connected to neither response, Response 1 or Response 2, respectively.

The operations, illustrated in Fig. 6.5, of applying attention weights to the locations of the letter items of a given display type can be summarized by expressing the attention weighting function defined across locations within a display as a vector,

$$\mathbf{w}_j = < \dots w_{-2}, w_{-1}, w_0, w_1, w_2, \dots >$$

and expressing the partitioning of elements associated with each item location by a matrix,

$$
\mathbf{C}_{jk} = \left\{ \begin{array}{ccc}
c_{-20} & c_{-21} & c_{-22} \\
c_{-10} & c_{-11} & c_{-12} \\
c_{00} & c_{01} & c_{02} \\
c_{10} & c_{11} & c_{12} \\
c_{20} & c_{21} & c_{22}
\end{array} \right\}
$$

where the row subscript, j, denotes the location of a letter item in a string of items; the value of the subscript represents the distance of an item from the target location at $j = 0$. The column subscript, k, denotes the type of response that elements in a set is connected to. Specifically, c_{j0} denotes a set of elements in a letter item connected to neither response, c_{j1} denotes a set of elements connected to Response type 1, and c_{j2} denotes a set of elements connected to Response type 2.

The actual operation of weighting each letter location in a display is given by the product

$$\mathbf{w}_j \mathbf{C}_{jk} = \mathbf{d}_k \text{ , where}$$

$$\mathbf{d}_k = \; < d_0, \, d_1, \, d_2 > \, ,$$

which may be termed the *weighted display vector*.

The notation of the display vector, $\mathbf{d}_k$, may be expanded to denote a particular one of the four types of displays used in the present experimental situation (Incompatible, Compatible, Same, or Neutral), by the subscript F, to $\mathbf{d}_{Fk}$,

$$F = I, \, C, \, S, \, N.$$

In addition, we can index the degree of attention spread applied to a given display by the subscript i, which was used for this purpose in the equation of attention spread (Equation (4)), so that the completely specified vector of the attention-weighted display is $\mathbf{d}_{Fki}$. It is this vector that represents the total set of elements available to the response counters that arise from the stimulus display.

But, in general, we should allow for the possibility that elements entering response counters may arise from sources other than the stimulus display. We call the group of elements from these other sources *background elements*. Some background elements arise from response biases, which we represent by set proportions β_1 and β_2, corresponding to element sets connected to Response type 1 and Response type 2, respectively. Response biases may be relatively constant over trials for a given subject, or may vary in an apparently random or spontaneous manner, due to momentary fluctuations of motor preparedness of one response or the other. These random motoric fluctuations presumably contribute to incorrect (wrong) responses. Other background elements are assumed to be associated with noise, and we assume that in this set the elements are neutral, and are denoted by β_0 (illustrated by the small 0's entering the response counters in Fig. 6.4).

Thus, the background element vector may be expressed as

$$\boldsymbol{\beta}_k = \; <\beta_0, \, \beta_1, \, \beta_2>,$$

such that

$$\sum_{k=0}^{2} \beta_k = 1.$$

The proportion of display elements relative to the total of display plus background elements is denoted by u. Then we can define

$$\mathbf{p}_{Fik} = u\mathbf{d}_{Fik} + (1 - u)\,\boldsymbol{\beta}_k.$$

We are now in a position to specify the summary vector of element proportions from both display and background sources that are available to the response counters on a trial as

$$\mathbf{p}_{Fik} = \langle p_{Fi0}, p_{Fi1}, p_{Fi2} \rangle,$$

such that

$$\sum_{k=0}^{2} p_{Fik} = 1.$$

It is this vector that is entered into the response probability equations and response time equations of the counter models to be described next.

Predictions of the Attention Theory Using Two Classes of Counter Models of Response Time

Given the vector expression $\mathbf{p}_{Fik}$, representing the particular mix of element types generated by a stimulus display and the background, we now consider two general classes of models that describe how a counting operation performed on a stream of elements eventually produces a given response, and at a particular time interval after the onset of the stimulus display. Both classes of counter models considered here assume that a response counter, illustrated in Fig. 6.4, accumulates elements of its own type until one counter first obtains its criterion number of elements, at which time a response of that type is emitted.

For one class of models, the class of Independent Counter Models, each response counter operates independently of the other counter. For the second class of models, the class of Interactive Counter Models, it is assumed that when an element increments one counter, the criterion count of the other counter is incremented by one unit. One could give a network interpretation of such an interaction as an inhibitory link between networks representing each response counter. To illustrate the Interactive Counter Model informally, one could insert a pair of reciprocal inhibitory connections between the two counters shown in Fig. 6.4.

During some small time unit, elements entering response counters can be assumed to enter them simultaneously, or to enter them only one at a time. The latter case serves as the basis of a serial discrete recruitment model (LaBerge, 1961; Luce, 1986) and a serial discrete random-walk model (LaBerge, 1990c). The former case serves as the basis of a parallel Poisson model (Townsend & Ashby, 1983), whose equations for response probability and response times can be shown to be equivalent to those of the recruitment model (LaBerge, 1990c).

The predicted mean response times for the various types of displays of the target-flanker task illustrated in Fig. 6.1 as a function of the attention weight at the target location are shown in Figs. 6.6 and 6.7. The attention weight at the target location represents the degree of attention concentration at that location. When the attention weight value is subtracted from 1.00, one obtains the weight given to all of the flanker locations, which indicates the amount of spread. The parameter r represents the criterion count for a particular response, that is the number of elements that a response counter needs to accumulate to produce a response. The attention weight is presumed to be varied experimentally by changes in the duration or size of Target 1 (see Fig. 6.2 and Table 6.1).

The predicted patterns of data shown in Figs. 6.6 and 6.7 can be compared with the obtained pattern of data shown in Fig. 6.3 and Table 6.1. Both the predicted curves and the empirical curves show that the Incompatible and Compatible response times converge as the attention weight of the target location increases (and the weight given to flankers decreases). The obtained increase in absolute response time with convergence is at present presumed to arise from nonattentional factors, and hence is not part of the predictions from the attention model. The two curves for the Neutral displays represent upper and lower bounds. The N_{100} curve denotes the case in which 100% of the elements of the Neutral displays are neutral with respect to the two response counters, that is, the flanker items of the Neutral displays bear zero similarity to the target items of the non-neutral displays. The N_0 curve denotes the maximum similarity case, in which all of the elements of a flanker item in a Neutral display are shared with the four Target 2 items. In this latter case, it is assumed that half of the elements of a flanker item are connected to one response counter and half to the other response counter.

The data for the Neutral display, shown in Fig. 6.3, and the data presented in the section titled "Attempts to Control Attention Spread" both appear to conform more closely to the predictions from the Interactive Counter (Random-Walk) Model than to the Independent Counter (Recruitment) Model. Further comparisons between the two types of response time models can be made on the basis of response time density functions, hazard functions, and error response times, but preliminary comparisons of these properties of the models (LaBerge, 1990c) indicate that the Neutral display data curve seems to offer the most conspicuous contrast found so far between the two types of response time models.

ACKNOWLEDGMENTS

This research was supported by the Office of Naval Research.

REFERENCES

Bjork, E. L., & Murray, J. T. (1977). On the nature of the input channels in visual processing. *Psychological Review, 84,* 472–484.

Eriksen, B. A., & Eriksen, C. W. (1974). Effects of noise letters upon the identification of a target letter in a nonsearch task. *Perception and Psychophysics, 16,* 143–149.

Estes, W. K. (1950). Toward a statistical theory of learning. *Psychological Review, 57,* 94–107.

Estes, W. K. (1972). Interactions of signal and background variables in visual processing. *Perception and Psychophysics, 12,* 278–286.

Estes, W. K., & Burke, C. J. (1953). A theory of stimulus variability in learning. *Psychological Review, 60,* 276-286.

Estes, W. K., & Straughan, J. H. (1954). Analysis of a verbal conditioning situation in terms of statistical learning theory. *Journal of Experimental Psychology, 47,* 225–234.

Kahneman, D., & Chajczyk, D. (1983). Tests of the automaticity of reading: Dilution of Stroop effects by color-irrelevant stimuli. *Journal of Experimental Psychology: Human Perception and Performance, 9,* 497–509.

LaBerge, D. (1959). A model with neutral elements. In R. R. Bush & W. K. Estes (Eds.), *Studies in mathematical learning theory* pp. 53–64. Stanford: Stanford University Press.

LaBerge, D. (1962). A recruitment theory of simple behavior. *Psychometrika, 27* 4 375–396.

LaBerge, D. (1983). The spatial extent of attention to letters and words. *Journal of Experimental Psychology: Human Perception and Performance, 9,* 371–379.

LaBerge, D. (1990a). Attention. *Psychological Science, 1*(3), 156–162.

LaBerge, D. (1990b). Thalamic and cortical mechanisms of attention suggested by recent positron emission tomographic experiments. *Journal of Cognitive Neuroscience, 2,* 358–372.

LaBerge, D. (1990c). *A quantitative model of attention and response processes in shape identification tasks.* (Technical Report MBS 90-18), Mathematical Behavioral Sciences, University of California, Irvine.

LaBerge, D., & Brown, V. (1989). Theory of attentional operations in shape identification. *Psychological Review, 96,* 101–124.

LaBerge, D., Brown, V., Carter, M., Hartley, A., & Bash, D. (1991). Reducing the effects of adjacent distractors by narrowing attention. *Journal of Experimental Psychology: Human Perception and Performance, 17.* 65–76.

Luce, R. D. (1986). *Response times.* New York: Oxford University Press.

Neimark, E. D., & Estes, W. K. (1967). *Stimulus sampling theory.* San Francisco: Holden-Day.

Shepard, R. (1989). Toward a universal law of generalization for psychological science. *Science, 237,* 1317–1323.

Townsend, J. T., & Ashby, F. G. (1983). *The stochastic modeling of elementary psychological processes.* New York: Cambridge University Press.

7 Triple Correlation and Texture Discrimination

John I. Yellott, Jr.
University of California, Irvine

INTRODUCTION

Over the past three decades a good deal of research on visual texture perception has been motivated by the hope that the discriminability of textures could be predicted on the basis of the global statistical properties of images. An early and especially influential expression of that hope was Bela Julesz's (1962) famous conjecture that textures cannot be discriminated without point-by-point scrutiny if they have the same "second-order statistics." Second-order statistics are descriptors applicable to images composed of a small number of distinct colors, and Julesz's conjecture takes its simplest form in the context of binary (black and white) images. The second-order statistics of a binary image are the probabilities, for all oriented line segments, that when any given segment is dropped at random onto the image (preserving its orientation), both ends land on the foreground color—for example, both on black. When this definition is formalized, the Julesz conjecture for binary textures proves to be equivalent to the hypothesis that two such textures cannot be discriminated ("preattentively," that is) if they have the same autocorrelation function. Figure 7.1 illustrates some perceptual facts that might lead one to such a hypothesis. On the left is an 8 × 8 array composed of four texture samples: The upper right quadrant is a 4 × 4 array of down-pointing arrows, and the other three quadrants are identical 4 × 4 arrays of up-pointing arrows. Considered as separate binary images, all four quadrants have the same autocorrelation function, and at first glance one does not notice that the upper right quadrant is different from the rest of the figure. In general, a 180° rotation of the micropatterns (here, the arrows) of a texture does not alter its autocorrelation function (for reasons spelled out in the section titled Autocorrela-

133

FIG. 7.1A. Textures composed of the same micropattern (here, the arrows) rotated by 180 degrees are not immediately discriminable. FIG. 7.1B. Ninety degree rotation of the same micropattern leads to immediate discrimination.

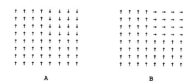

tion and Second-Order Statistics), and it appears to be generally true that we cannot instantaneously discriminate between binary textures composed of micropatterns that are 180° rotations of one another. This fact gains significance when one realizes that the same is not true of 90° rotations. The right side of Fig. 7.1 shows another 4 × 4 arrangement of texture samples, but here the upper right quadrant is composed of right-pointing arrows. In this case the autocorrelation function of the upper right quadrant image is not the same as that of the other three quadrants, and perceptually, that quadrant is immediately seen to be different.

Thus texture pairs composed of micropatterns that are identical apart from a 180° degree rotation provide supporting evidence for Julesz's second-order statistical conjecture. During the 1960s and early 1970s a great deal of ingenuity was devoted to finding other ways of constructing binary textures that are physically distinct but have the same second-order statistics, and initially the Julesz conjecture held up remarkably well (Julesz, 1975; Julesz, Gilbert, Shepp, & Frisch, 1973). However, by the early 1980s it seemed to have been decisively refuted, first by Julesz, Gilbert, and Victor (1978), and again by Diaconis and Freedman (1981). Both papers presented pairs of binary textures (like those in Fig. 7.2) that are immediately discriminable, but that were said to have identical "third-order statistics." The third-order statistics of a binary image are the probabilities, for all triangles, that when any given triangle is dropped at random onto the image (preserving its orientation), all three corners land on black points. Because images that have identical third-order statistics must have identical second-order statistics (line segments being degenerate triangles), the images of Julesz et al. (1978) and Diaconis and Freedman, seemed to clearly contradict Julesz's second-order conjecture. Moreover, they appeared to bar any possible extension of that conjecture to third-order statistics. These apparently decisive counterexamples led researchers to reject simple correlational approaches to texture discrimination, and to explore new theoretical directions, such as Julesz's (1981) own "texton" theory.

However, the demise of the Julesz conjecture now appears to have been reported prematurely. Recently Geoffrey Iverson and I (Yellott & Iverson, 1989) have been studying a generalized form of autocorrelation that was introduced to the texture perception literature in 1986 by Klein and Tyler. (Actually, unbeknownst to Klein and Tyler, and initially to me and Iverson, physicists had been exploiting the same ideas for some time. Lohmann and Wirnitzer, 1984, provide

FIG. 7.2. Binary texture-pairs created according to the algorithms of Julesz, Gilbert, and Victor (1978) (top figure), and Diaconis and Freedman (1981) (bottom figure). The top figure is two 50 × 50 pixel images side by side; tick marks at top and bottom indicate the break between them. The left image is created by coloring the pixels in its left column and bottom row randomly black or white with .5 probability. The remaining pixels are then colored according to the rule that every four adjacent pixels must contain an even number of black pixels. The right image is created by coloring the left column and bottom row pixels randomly black or white (here, using the same random coloring as in the left image), and then coloring the remaining pixels according to the rule the every four adjacent pixels must contain an odd number of blacks. Julesz et al. claimed that these algorithms produce images that have the same third-order statistics, but triple correlation theory shows that cannot be true. The bottom figure is again two 50 × 50 pixel images side by side. The left image is created by randomly coloring each pixel in the bottom row black or white with .5 probability. Then each of the remaining rows is colored the same as the bottom row, or the opposite, with probability .5. The right image is created by coloring each pixel randomly black or white, with probability .5 for all pixels. Diaconis and Freedman claimed that those two algorithms produce images having the same third-order statistics; again, triple correlation theory shows this is impossible.

135

an introduction to the physics literature.) Our results show that it is actually mathematically impossible to create pairs of discriminable binary images that have the same third-order statistics, because any two images that have identical third-order statistics must be physically identical. This conclusion follows from the fact that binary images which have the same third-order statistics must have the same "triple correlation" function (in the same way that binary images with identical second-order statistics have the same autocorrelation function), and any image of finite size is uniquely determined, up to a translation, by its triple correlation. Consequently two black and white texture samples that have the same third-order statistics must be physically identical except for their positions on the page, and discrimination between them is impossible by definition.

Thus the image-pairs presented by Julesz et al. (1978) and Diaconis and Freedman (1981) did not, in fact, demonstrate that people can discriminate between textures that have the same third-order statistics: Triple correlation theory shows that such a discrimination is impossible in principle. Moreover, empirical analysis of the images from both papers shows that neither pair had identical second-order statistics. Consequently they also did not represent counterexamples to Julesz's original second-order conjecture.

These conclusions raise two questions. First, what was the actual statistical relationship between the image-pairs of Julesz et al. (1978) and Diaconis and Freedman (1981)? Second, are there, in fact, pairs of black and white textures that are immediately discriminable despite having identical second-order statistics?

The answer to the first question is the same for both Julesz et al. (1978) and Diaconis and Freedman (1981). What both papers actually provided were not pairs of specific images with identical third-order statistics, but pairs of probabilistic image-generating algorithms (i.e., 2-D stochastic processes) whose third-order ensemble statistics are identical. That is, across all the images that will be generated (over time) by two algorithms, the probability that any fixed triple of points (say, the points [1,2], [3,4], [5,6]) will all be black is the same for both algorithms. Such a probability is not a property of any specific physical image that might be viewed by an observer. Instead, it is a property of the infinite set of images that an algorithm would generate if it were repeated infinitely often—a property of the movie that the algorithm would create, but not of any single frame. Thus it is not a property that should be expected to determine the perceptual response of an observer viewing a single image.

As to the second question: The literature does contain other examples of binary textures that are said to violate Julesz's second-order conjecture (e.g., Julesz, 1981, p. 94, Fig. 5). However, as far as I can determine, none of the published counterexamples actually involve discriminable texture pairs that have identical second-order statistics. Instead, in all cases, the technique for creating second-order identity involves a random rotation of micropatterns that themselves have different second-order statistics; the idea being that infinitely large arrays of such rotated micropatterns will have the same global second-order statistics. But for the finite-size images that can be viewed by an observer, this technique cannot be expected to create pairs of images that actually have identical second-order statistics. Thus it appears that the literature does not currently provide any decisive counterexample to Julesz's conjecture that binary textures with identical second-order statistics cannot be discriminated preattentively.

To remedy this, I describe here a new technique for creating texture pairs whose second-order statistics are strictly identical, but that are immediately discriminable. Thus Julesz's second-order conjecture, whose demise was prematurely announced a decade ago, can now be regarded as safely dead.

The next section reviews some facts about ordinary autocorrelation, and explains the relationship between the autocorrelation function of an image and its second-order statistics. The titled "Triple Correlation and Third-Order Statistics" section introduces the triple correlation, indicates how one proves that every image of finite size is uniquely determined up to translation by its triple correlation function, and explains the relationship between triple correlation and third-order image statistics. It follows immediately from that relationship that one cannot construct physically different binary images that have identical third-order statistics. Finally, the last section discusses the second-order statistical properties of the textures presented by Julesz et al. (1978) and Diaconis and Freedman (1981), and describes the new method for creating textures that are immediately discriminable despite having identical second-order statistics.

AUTOCORRELATION AND SECOND-ORDER STATISTICS

Suppose $f(x,y)$ is a real-valued nonnegative function with finite support, representing the luminance at each point (x,y) in a monochromatic image of finite size. To construct the ordinary autocorrelation function of this image we create a shifted copy of f, of the form $f(x + s, y + t)$, multiply that copy times the original image, and integrate the product over all x,y values. If we do this for all possible shifts we have a new function $a_f(s,t)$—a function of the shift arguments s and t. This is the autocorrelation of the image f. Formally then, the autocorrelation function of f is defined by

$$a_f(s, t) = \int_{-\infty}^{\infty} \int_{-\infty}^{\infty} f(x, y) \, f(x + s, y + t) \, dxdy. \tag{1}$$

To see how the autocorrelation a_f is related to f itself, we calculate its Fourier transform $A_f(u, v)$, and find that

$$A_f(u, v) = |F(u, v)|^2 \tag{2}$$

where F(u, v) is the Fourier transform of f, that is,

$$F(u, v) = \int_{-\infty}^{\infty} \int_{-\infty}^{\infty} e^{-i2\pi(ux+vy)} f(x, y) \, dxdy \qquad (3)$$

and $|F(u,v)|^2$ is the square modulus of F, which (because f is real) equals $F(u,v)F(-u,-v)$. The Fourier transform $F(u,v)$ is a complex-valued function of the real arguments u,v, which can be expressed in complex exponential form as $|F(u,v)|e^{iPhaF(u,v)}$, where $|F(u,v)|$ is a nonnegative real function called the *amplitude spectrum* of f, and $PhaF(u,v)$ is another real function called the *phase spectrum*. The fundamental inversion theorem of Fourier analysis shows that f is uniquely specified by its phase and amplitude spectra via the relationship

$$f(x, y) = \int_{-\infty}^{\infty} \int_{-\infty}^{\infty} |F(u, v)| \cos[2\pi(ux + vy) + PhaF(u, v)] \, dudv$$

which essentially means that f is the sum over all u,v of (co)sinusoidal ("spatial frequency") components whose amplitude for each u,v pair is given by $|F(u,v)|$ and whose phase is given by $PhaF(u,v)$. Thus the autocorrelation of a_f of an image f tells us the amplitudes of all the spatial frequency components of that image, but none of their phases. In effect, then, it tells us half of what we need to know to specify the image.

It is easy to construct images that look very different but have the same autocorrelation function. The obvious way is to decompose any image into its Fourier components, randomize the phases of those components while leaving their amplitudes the same, and invert the resulting transform. When this operation is performed on a real world scene, the result will generally be quite unrecognizable. But for the purpose of testing the Julesz conjecture, this kind of phase scrambling operation is not useful, because when it is performed on a binary texture image, the result will almost surely not be another binary image: the blacks and whites of the original will be smeared into shades of grey.

In fact, it is surprisingly difficult to find ways of creating pairs of distinct binary images that have identical autocorrelations. As far as I know, the texture perception literature has relied on only one basic trick. (That is, for producing images with identical autocorrelations. Statistical methods for creating binary images with roughly similar autocorrelations are more common. These are discussed in the last section.) This is to exploit the fact that turning an image upside down leaves its autocorrelation function unchanged. If f(x,y) describes an image, $f(-x,-y)$ is that image rotated by 180°. The Fourier transform of $f(-x,-y)$ is $F(-u,-v)$, so the power spectrum of $f(-x,-y)$ is the same as that of f(x,y), that is, $|F(u,v)|^2$. Thus, rotating an image by 180° has no effect on its autocorrelation.

In general, of course, images look quite different when turned upside down. Thus, as Julesz rightly stressed, it is quite remarkable that a 180° rotation of the micropatterns of a texture seems to go unnoticed by the "preattentive" visual

FIG. 7.3. Images A, B, C, and D all have the same autocorrelation function, as explained in the text. The bottom part of the figure shows the same four images slid together. At a glance, they appear to form a single texture.

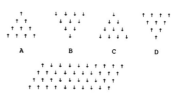

system. As noted earlier, this fact becomes more striking when one realizes that the same is not true of 90° rotations. Ninety-degree rotation of an image does not preserve its autocorrelation, because it changes $f(x,y)$ into $f(-y,x)$ or $f(y,-x)$, making the power spectrum $F(v,-u)F(-v,u)$, which in general does not equal $|F(u,v)|^2$. Observations of this sort suggest that texture discrimination might be based on visual mechanisms that could be characterized by simple rules—such as the one embodied in the Julesz conjecture.

Before turning to that conjecture, I should deal with a mathematical point that may be bothering the reader. If $f(x,y)$ describes a micropattern, such as an up-pointing arrow, we have seen that rotating f by 180° into $f(-x,-y)$ leaves its autocorrelation function unchanged. But suppose we have an image composed of multiple copies of the micropattern f placed at various locations in the plane—a texture, in other words. This texture then is represented by a function of the form $m(x,y) = f(x - x_1, y - y_1) + f(x - x_2, y - y_2) + ...,$ where the coordinates (x_i,y_i) give the location of the ith copy of f. We know that turning any image upside down leaves its autocorrelation function unchanged, so if we turn the entire texture $m(x,y)$ upside down, into $m(-x,-y)$, that has no effect on its autocorrelation. However this is not equivalent to turning the micropattern $f(x,y)$ upside down, and then placing copies of the rotated micropattern at the locations (x_i,y_i). Is it true that the latter operation also leaves the autocorrelation unchanged? Figure 7.3 illustrates this point. Suppose $m(x,y)$ is an up-pointing pyramid composed of up-pointing arrows, as in Fig. 7.3A. Then $m(-x,-y)$ is a down-pointing pyramid composed of down-pointing arrows, so image B in Fig. 7.3 has the same autocorrelation as A. Rotating the micropatterns within $m(x,y)$ yields an up-pointing pyramid composed of down-pointing arrows: Fig. 7.3C. The question is, does C have the same autocorrelation as A and B?

The answer is yes, for the following reason. We can represent the texture $m(x,y)$ as the convolution of the micropattern $f(x,y)$ and a function consisting of Dirac delta functions at the locations (x_i,y_i), that is, a function of the form $a(x,y) = \Sigma_i \delta(x - x_i, y - y_i)$. Thus $m(x,y) = f(x,y)*a(x,y)$. This implies that in the Fourier transform domain we have $M(u,v) = F(u,v)A(u,v)$, where M and A are the transforms of m and a, and thus the power spectrum of the texture is $|M(u,v)|^2 = |F(u,v)|^2|A(u,v)|^2$. Now if we rotate the entire texture into $m(-x,-y)$, its power spectrum is still $|M(u,v)|^2$. But if we rotate the micropattern f into $f(-x,-y)$, and convolve this with $a(x,y)$, the power spectrum of the resulting texture f(-x,-y)*a(x,y) is F(-u,-v)F(u,v)A(u,v)A(-u,-v), which also equals $|M(u,v)|^2$. So the autocorrelation function of f(-x,-y)*a(x,y) is indeed the same as that of the original texture m(x,y) and the rotated texture m(-x,-y). One can see that this is also true of f(x,y)*a(-x,-y), so Fig. 7.3D has the same autocorrelation function as A, B, and C.

Now for second-order statistics. Suppose f(x,y) describes a binary image, that is, f(x,y) = 1 if the point (x,y) has the foreground color (say, black), and f(x,y) = 0 if (x,y) has the background color (white). Julesz (1981, p. 91) defined the second-order statistics of such an image to be the probabilities "of the events that the vertices (end points) of randomly thrown 2gons (dipoles or needles) of all possible lengths and orientations will fall on certain colors of the texture (for example, both on black)." To formalize this definition, we imagine that the image is contained in some finite region of the plane, say inside a square region of area R centered at the origin. Then any given dipole (i.e., any line segment of some fixed length and orientation) corresponds to a vector of the form (s,t) (i.e., the line from [0,0] to [s,t]). Dropping this vector at random onto the image means picking a point (x,y) at random inside the image region for one end of the vector; then (x + s, y + t) is its other end. We seek the probability that both (x,y) and (x + s, y + t) are black, that is, that f(x,y) and f(x + s, y + t) both equal 1, which we will denote by $j_{2,f}(s,t)$. Let (**x,y**) be a 2-dimensional random vector uniformly distributed over the image region, so that the probability density function of (**x,y**) is 1/R for all (x,y) in the image region and zero elsewhere. Then f(**x,y**) is a random variable that equals 1 if the random point (**x,y**) is black and equals zero otherwise, and for any fixed values of s and t the product f(**x,y**)f(**x** + s, **y** + t) is a random variable that equals 1 if the points (**x,y**) and (**x** + s, **y** + t) are both black, and equals zero otherwise. So the second-order statistic for the vector (s,t) is the probability that the product random variable f(**x,y**)f(**x** + s, **y** + t) equals 1, which is the same as the expected value of that random variable, that is,

$$j_{2,f}(s, t) = (1/R) \int_{-\infty}^{\infty} \int_{-\infty}^{\infty} f(x, y)f(x + s, y + t) \, dxdy. \qquad (4)$$

Comparing (4) with (1), we see that

$$j_{2,f}(s,t) = (1/R)a_f(s,t), \qquad (5)$$

that is, the second-order statistic function of a binary image equals the autocorrelation function of that image divided by its area. (We note that the same result will be obtained if we interpret dropping an oriented needle "at random" to mean that its centerpoint is uniformly distributed over the image area.)

A little manipulation of Equation 5 [bearing in mind that here, because f = 1 or 0, $f(x,y)^2 = f(x,y)$] quickly shows that the image area R equals $J_{2,f}(0,0)/[j_{2,f}(0,0)^2]$, where $J_{2,f}$ is the Fourier transform of $j_{2,f}$. So the second-order statistic function of any binary image determines the autocorrelation function of that image, and two images that have identical second-order statistics must have identical autocorrelations. Conversely, if two binary images have identical autocorrelation functions and the same size, they must have identical second-order statistics. Thus, for example, the nondiscriminable arrays of equally spaced up-pointing or down-pointing triangles in Fig. 7.1A have the same second-order statistics.

As an aside, for non-binary images, Julesz's second-order statistics are not equivalent to the autocorrelation function. If an image contains several different luminance levels, say l_1, l_2, ..., l_n, then its second-order statistics are the probabilities, for all luminance combinations l_i, l_j, and all line segments (s,t), that when segment (s,t) is randomly dropped on the image one end lands on luminance l_i and the other on l_j. In this case, two images can have the same autocorrelation function, but different second-order statistics. It is easy to construct texture pairs containing three luminance levels that are immediately discriminable and have the same autocorrelation function (Julesz, 1962, p. 88, Fig. 9 is an example), but these are not necessarily counterexamples to the Julesz conjecture, which requires all the second-order statistics of the images to be identical. Constructing distinct non-binary images with identical second-order statistics is a complicated matter, which is why Julesz's conjecture has been studied almost exclusively in the context of black and white images.

Triple Correlation and Third-Order Statistics

The autocorrelation operation defined by Equation 1 lends itself to an obvious generalization: Instead of integrating the product of the image f(x,y) and a single shifted copy $f(x + s, y + t)$, one could create two shifted copies, $f(x + s_1, y + t_1)$ and $f(x + s_2, y + t_2)$, form the product of all three images, and integrate that product over all x,y. The new function $t_f(s_1, t_1, s_2, t_2)$ created in this way has come to be called the "triple correlation" of the image f (e.g., by Lohmann & Wirnitzer, 1984); its formal definition is

$$t_f(s_1, t_1, s_2, t_2) = \int_{-\infty}^{\infty} \int_{-\infty}^{\infty} f(x, y) f(x + s_1, y + t_1) f(x + s_2, y + t_2) \, dxdy. \tag{6}$$

(This definition, like the earlier definition of the autocorrelation function given by Eq. 1, assumes that f itself is an integrable function. We are concerned with images of finite size, so that assumption is satisfied naturally.)

At this point it will be helpful to streamline our notation. Hereafter X denotes a 2-D vector (x,y); f(X) denotes the image f(x,y); and S_i denotes a shift vector (s_i, t_i). Then definition (6) can be written compactly as

$$t_f(S_1, S_2) = \int_{-\infty}^{\infty} \int_{-\infty}^{\infty} f(X) \, f(X + S_1) \, f(X + S_2) \, dX. \tag{7}$$

Now to see what the triple correlation t_f of an image f tells us about f, we take its Fourier transform T_f, which in vector notation is

$$T_f(U_1, U_2) = \int_{-\infty}^{\infty} \int_{-\infty}^{\infty} \int_{-\infty}^{\infty} \int_{-\infty}^{\infty} \exp[-i2\pi(U_1 \cdot S_1 + U_2 \cdot S_2)] \, t_f(S_1, S_2) \, dS_1 dS_2$$

where $U_i = (u_i, v_i)$ and $U_i \cdot S_i = (u_i s_i + v_i t_i)$. In the physics literature, the transform T_f is called the "bispectrum" of f. A straightforward calculation shows that it satisfies the following elegant relationship:

$$T_f(U_1, U_2) = F(U_1) \, F(U_2) \, F(-U_1 - U_2) \tag{8}$$

where, as before, $F(U_i) = F(u_i, v_i)$ is the Fourier transform of f. Setting $U_2 = -U_1$ in Equation 8 we see that $T_f(U, -U) = F(0)|F(U)|^2$, showing that the bispectrum of an image determines its power spectrum—and thus, the triple correlation determines the autocorrelation. Equation 8 also shows immediately that translating an image, say from f(x,y) to f(x + a, y + b), leaves its triple correlation unchanged: If g(x,y) = f(x + a, y + b), its transform G(u,v) is $e^{i2\pi(ua + vb)}F(u,v)$, so that in the bispectrum expression for T_g the exponential phase factors cancel, leaving $T_g(U_1, U_2) = T_f(U_1, U_2)$.

Finally, and most importantly, the bispectrum Equation 8 can be used to show that any image of finite size is uniquely determined, up to translation, by its triple correlation. In other words, if two pictures have the same triple correlation, then they must be identical pictures, differing at most in where they hang on the wall. (Note that this is true of all finite images, not simply binary ones.)

I am aware of two ways to prove this result, both unfortunately indirect and rather lengthy. The proof commonly cited in the physics literature is due to Bartelt, Lohmann, and Wirnitzer (1984); it relies on the fact that the complex Fourier transform of a function with finite support is uniquely specified (up to a translation factor) by its zeros, and those zeros are determined by the zeros of its bispectrum. Iverson and I have devised a different proof, which I sketch here. It has the technical advantage of being extendable to certain functions that do not have finite support, and consequently cannot be handled by the proof of Bartelt et al. (1984) because the Fourier transforms of such functions need not have any zeros. (Gaussians are an example.)

Here is the theorem we wish to prove: **If f(X) and g(X) are nonnegative real functions with finite support, then $t_f(S_1, S_2) = t_g(S_1, S_2)$ for all S_1, S_2 if and only if g(X) = f(X + C) for some constant C = (c_1, c_2).** The "if" part is obvious, since the triple correlation is unaffected by a translation. To show "only if," we start with the fact that if $t_f = t_g$, then in the spectral domain the bispectrum relationship (Equation 8) implies

$$F(P) \, F(Q) \, F(-P - Q) = G(P) \, G(Q) \, G(-P - Q) \tag{9}$$

for all P = (p_1, p_2), Q = (q_1, q_2). Setting P = Q = 0, that is, (0,0), Equation 9 implies F(0) = G(0). This means that f and g have the same integral over the whole plane. If that integral is zero, f and g must themselves be zero, and thus trivially equal to one another, except on a set of measure zero. In that case the images f and g are both physically indistinguishable from a totally blank image, because their integrals over any region—the total measurable amount of light in any region—must be zero. So we can assume $F(0)$ and $G(0)$ are nonzero, and without loss of generality we can assume their common value is 1.0. (If it is not, we can divide t_f by $F(0)^3$ and t_g by $G(0)^3$, and deal with the normalized functions $f(X)/F(0)$ and $g(X)/G(0)$.) In that case f and g are 2-dimensional probability density functions, and F and G are the characteristic functions of those densities. Because f and g have finite support, both densities are completely determined by the values of their characteristic functions in a neighborhood of the origin. This follows from the well-known fact (e.g., Feller, 1966, ch. XV) that a one-dimensional probability density function with finite support is completely specified by the derivatives of its characteristic function at zero, which determine the whole characteristic function via a Taylor series expansion. That result can be quickly generalized to two dimensions: If $f(x,y)$ is the density function of a random vector $(\mathbf{x},\mathbf{y})$, with characteristic function $F(u,v) = E[e^{-i2\pi(ux+vy)}]$, and f has finite support in the plane, then for every fixed pair of u,v values, say u',v', the sum $u'\mathbf{x} + v'\mathbf{y}$ is a one-dimensional random variable whose density has finite support on the line. Thus the characteristic function of $u'\mathbf{x} + v'\mathbf{y}$, say $c_{u'v'}(t) = E[e^{-i2\pi t(u'\mathbf{x}+v'\mathbf{y})}]$, is completely determined for all t by its values in a neighborhood of $t = 0$, which of course are values of $F(u,v)$ in a neighborhood of $(0,0)$. In particular this is true of $c_{u'v'}(1)$, which is $F(u',v')$. So for every pair of u,v values, $F(u,v)$ is determined by the values of F in a neighborhood of the origin.

Now to complete the proof, we now show that Equation 9 implies that there must be a vector constant $C = (c_1,c_2)$ such that for all P in some neighborhood of the origin (and thus, for all P generally, from the argument given), $G(P) = e^{i2\pi C \cdot P}F(P)$. This in turn implies that $g(X) = f(X + C)$. To show this, we appeal to the fact that all characteristic functions are continuous (e.g., Feller, 1966, ch. XV), and we know that $F(0) = G(0) = 1$, so there is a neighborhood of the origin in which both F and G are nonvanishing and can be divided freely. Dividing appropriately in Equation 9, and using the facts that $1/G(-P-Q) = G(P + Q)/|G(P + Q)|^2$ and $F(-P-Q) = |F(P + Q)|^2/F(P + Q)$ (because F and g are transforms of real functions), and $|F(P + Q)|^2 = |G(P + Q)|^2$ (because setting Q $= -P$ in Equation 9 implies $|F(P)|^2 = |G(P)|^2$ for all P), we obtain the following functional equation for the function $H(P) = G(P)/F(P)$

$$H(P + Q) = H(P)H(Q), \qquad (10)$$

which is valid for all P and Q in some neighborhood of the origin. Equation 10 is a complex version of the well-known Cauchy functional equation $f(x + y) = f(x)f(y)$, whose only continuous solution for real f is $f(x) = e^{cx}$. In our case H is continuous (because G and F are), and nonvanishing in a neighborhood of the origin. Aczel (1966, pp. 213–216) shows that all continuous nonvanishing solutions of Equation 10 take the form $H(P) = \exp(B \cdot P)\exp(iC \cdot P)$, where B and C are vector constants, and because here $|H(P)| = |G(P)|/|F(P)| = 1$, the constant B must be 0. Consequently for some constant C, $G(P) = e^{iC \cdot P}F(P)$ in a neighborhood of the origin, hence for all P, and $g(X) = f(X + C)$ as claimed.

Now to apply this result to the third-order statistics of images. As in the previous section, let f(X) represent a binary image, with $f(X) = 1$ if point X has the foreground color (say, black) and $f(X) = 0$ if X has the background color. We assume the image has finite size, so the function f has finite support. The third-order statistic function of f, $j_{3,f}(S_1,S_2)$ is the probability that when the triangle defined by the points 0, S_1, and S_2 (i.e., the points (0,0), (s_1,t_1), and (s_2,t_2)) is dropped at random onto the image (without changing its orientation), all three corners fall on black points. To formalize this definition, let R be the area of the region of the plane containing the image f, and $X = (x,y)$ a random vector uniformly distributed on that region, with density 1/R. Then $j_{3,f}(S_1, S_2)$ is the probability that $f(X)f(X + S_1)f(X + S_2)$ equals 1, which is the expected value of that product, so

$$j_{3,f}(S_1, S_2) = (1/R) \int_{-\infty}^{\infty} \int_{-\infty}^{\infty} f(X) \, f(X + S_1) \, f(X + S_2) \, dX. \qquad (11)$$

Comparing Equation 11 to Equation 7, we see that the third-order statistic function of a binary image is the triple correlation function of that image divided by the image area:

$$j_{3,f}(S_1,S_2) = (1/R) \, t_f(S_1,S_2). \qquad (12)$$

Equation 11 shows that $j_{3,f}(S_1,0) = j_{2,f}(S_1)$ (because here $f^2 = f$), and we showed earlier that the image area R is determined by $j_{2,f}$. Consequently if two binary images have the same third-order statistics they must have the same triple correlation. It follows that two binary images of finite size can have the same third-order statistics if and only if those images are identical up to a translation.

It was noted at the end of the last section that for nonbinary images the autocorrelation function is not equivalent to the second-order statistics, so that in general two images can have the same autocorrelation but different second-order statistics. For the triple correlation the situation is quite different: Any two monochromatic images of finite size (binary or otherwise) that have the same triple correlation must be physically identical up to translation, and thus necessarily must have the same third-order statistics.

Autocorrelation and Texture Discriminability

The counterexamples to the Julesz's conjecture presented in the papers of Julesz et al. (1978) and Diaconis and Freedman (1981) were motivated by the fact that binary images that have the same third-order statistics must have the same second-order statistics. Thus if one could construct pairs of immediately discriminable texture images that had identical third-order statistics, those pairs would

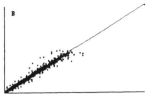

FIG. 7.4.   Scatterplots showing the relationship between the autocorrelation functions of the image-pairs in Fig. 7.2. Graph A shows values
of the autocorrelation function $a_f(s,t)$ of the image at the top left of Fig.
7.2 (created by the "even" algorithm of Julesz, Gilbert, and Victor,
1978) plotted against the corresponding values (i.e., for the same s,t
pairs) of the autocorrelation function of the image at the top right
(created by Julesz et al.'s "odd" algorithm). Points are shown for all 0
$\leq$ s,t $\leq$ 49; both axes run from 0 to 2500. All points should fall on the
diagonal line if the two images had the same autocorrelation function.
Graph B shows the same plot for the pair of images at the bottom of
Fig. 7.2, created by the algorithms of Diaconis and Freedman (1981).

necessarily have identical second-order statistics as well, and would constitute
counterexamples to Julesz's conjecture that textures with the same second-order
statistics cannot be discriminated without point-by-point scrutiny. The triple correlation uniqueness theorem proved in the last section shows that strategy for
disproving the Julesz conjecture is hopeless, because it is impossible to construct
images with identical third-order statistics that are not physically identical, and
thus indiscriminable by definition. But this does not necessarily imply that the
image-pairs presented in the two papers failed to disprove the Julesz conjecture,
because it is not impossible that they actually had identical second-order statistics, though not for the reasons originally claimed. To check this possibility I
have computed the autocorrelation functions of images created with the algorithms of Julesz et al. and Diaconis and Freedman. (Those algorithms are
described in the caption to Fig. 7.2.) As one might expect, the autocorrelation
functions are never identical. Figure 7.4 shows typical results for both pairs of
algorithms, using as examples the texture pairs shown in Fig. 7.2. The left side
shows values of the autocorrelation function of a 50 × 50 pixel image created
with Julesz et al.'s "even" algorithm (the lefthand texture in the top half of Fig.
7.2) plotted against the values of the autocorrelation function of a 50 × 50 image
created with their "odd" algorithm (the righthand texture in the top half of Fig.
7.2). It can be seen that autocorrelation functions are quite similiar, but clearly
not identical. The right side of the figure illustrates the same point for the
algorithms of Diaconis and Freedman. Tests with images of different sizes show
that the autocorrelation functions become increasingly similar as size increases,
and for infinite-size images we would expect them to all become identical (a

consequence of ergodicity). But infinite images cannot be shown to observers, and for finite images, the algorithms of Julesz et al. and Diaconis and Freedman do not produce image pairs with identical autocorrelations. So the easy discriminability of images like those in Fig. 7.2 does not disprove Julesz's second-order conjecture. Instead, it shows that one can immediately discriminate between binary textures whose second-order statistics are similar, though not identical.

The same can be said of the other counterexamples to the Julesz conjecture that one finds in the literature (illustrated in Julesz, 1981). Those counterexamples consist of binary texture pairs in which one texture is composed of a fundamental micropattern, replicated many times over the plane, each time in a random orientation, and the other texture is composed of a different micropattern, also replicated many times with each copy randomly rotated. The individual micropatterns from one texture do not have the same autocorrelation functions as those of the other texture, but because of the random rotation (and the ingenious construction of the micropatterns), the autocorrelation function of an infinite sample of one texture will have the same autocorrelation function as an infinite sample of the other texture. With this technique one can create pairs of binary texture images that are immediately discriminable (e.g., Figs. 5.A and 5.B in Julesz, 1981). However because the autocorrelation identity holds only for infinite images, and not for the small arrays of micropatterns that one can actually view, such images do not prove that observers can preattentively discriminate binary textures that have identical second-order statistics. Instead, like the examples of Julesz et al. (1978) and Diaconis and Freedman (1981), they only show that certain texture images can be immediately discriminated even though their second order statistics are roughly the same. In other words, they provide sug-

FIG. 7.5.   Counterexample disproving the Julesz conjecture. The top of the figure shows a pair of binary texture images that have identical autocorrelation functions, and that are immediately discriminable. Tick marks above and below indicate the break between the textures. The lefthand texture consists of three replications (compressed horizontally and stretched vertically) of the micropattern shown in the bottom left of the figure. The righthand texture consists of three replications of the micropattern shown at the bottom right. The construction principle is explained in the text.

gestive evidence that the Julesz conjecture is false, but they do not prove that it is.

A conjecture that has played such a central role in the history of texture perception research deserves a decisive refutation, especially since textbooks (e.g., Resnikoff, 1989) now treat it as having been long since disproved. Figure 7.5 provides a decisive counterexample to the Julesz conjecture. It shows a pair of immediately discriminable binary textures that have exactly identical autocorrelation functions, and thus identical second-order statistics. Each image consists of three replications of the micropatterns shown at the bottom of the figure, which have identical autocorrelation functions. (The micropatterns have been expanded horizontally for clarity, and compressed vertically for convenience.) The principle of their construction is as follows. Suppose $f(x)$ and $g(x)$ are any two real functions of a real variable. The Fourier transforms of the convolutions $f(x)*g(x)$ and $f(x)*g(-x)$ are $F(u)G(v)$ and $F(u)G(-u)$, so the power spectra of $f(x)*g(x)$ and $f(x)*g(-x)$ both equal $|F(u)|^2|G(u)|^2$. Thus $f(x)*g(x)$ and $f(x)*g(-x)$ have the same autocorrelation function. Now if $h(y)$ is any non-negative function, $h(y)[f(x)*g(x)]$ and $h(y)[f(x)*g(-x)]$ represent 2-dimensional images with identical autocorrelation functions, because both have a power spectrum equal to $|H(v)|^2|F(u)|^2|G(u)|^2$. To ensure that these images are binary, suppose $f(x)*g(x)$ and $f(x)*g(-x)$ are both composed entirely of Dirac deltas, that is,

$$f(x)*g(x) = \sum_{i=1}^{n} \delta(x - p_i)$$

for some set of distinct points $p_1, \ldots, p_n$, and likewise

$$f(x)*g(-x) = \sum_{i=1}^{n} \delta(x - q_i)$$

for distinct points $q_1, \ldots, q_n$; and let $h(y)$ be a function that equals 1 for $|y| <$ any value w, and equals zero elsewhere. Then the images $h(y)[f(x)*g(x)]$ and $h(y)[f(x)*g(-x)]$ will both consist of a row of vertical lines of equal length w and equal luminance; that is, both will be binary images. The micropatterns shown in Fig. 7.5 were constructed on this basis, using the functions

$$f(x) = \sum_{n=0}^{4} \delta(x - [2n]^2)$$

and

$$g(x) = \sum_{n=0}^{4} \delta(x - [2n]^2 - [n + 1]).$$

A numerical check shows that in this case $f(x)*g(x)$ and $f(x)*g(-x)$ are both binary, so the micropatterns $h(y)[f(x)*g(x)]$ and $h(y)[f(x)*g(-x)]$ have the same autocorrelation function. That equality is inherited by images composed of multiple replicas of these micropatterns, so the texture images on the left and right sides of the top part of Fig. 7.5 have identical autocorrelations. Since these binary texture images are immediately discriminable despite having identical second-order statistics, the Julesz conjecture is disproved.

ACKNOWLEDGMENTS

This chapter is dedicated to William K. Estes, with thanks for his teaching during my graduate school days and his help throughout my career.

I also thank Geoffrey Iverson for his collaboration on the theory of generalized autocorrelation functions, and Ram Kakarala for valuable discussions on texture perception.

REFERENCES

Aczel, J. (1966). *Lectures on functional equations*. New York: Academic Press.

Bartelt, H., Lohmann, A. W., & Wirnitzer, B. (1984). Phase and amplitude recovery from bispectra. *Applied Optics, 23*, 3121–3129.

Diaconis, P., & Freedman, D. (1981). On the statistics of the Julesz conjecture. *Journal of Mathematical Psychology, 24*, 112–138.

Feller, W. (1966). *An introduction to probability theory and its applications: Volume II*. New York: John Wiley & Sons.

Julesz, B. (1962). Visual pattern discrimination. *IRE transactions on information theory, IT-8*, 84–92.

Julesz, B. (1975). Experiments in the visual perception of texture. *Scientific American, 232*, 34–43.

Julesz, B. (1981). Textons, the elements of texture perception and their interactions. *Nature, 290*, 91–97.

Julesz, B., Gilbert, E. N., Shepp, L. A., & Frisch, H. L. (1973). Inability of humans to discriminate between textures that agree in second-order statistics—Revisited. *Perception, 2*, 391–405.

Julesz, B., Gilbert, E. N., & Victor, J. D. (1978). Visual discrimination of textures with identical second order statistics. *Biological Cybernetics, 31*, 137–140.

Klein, S. A., & Tyler, C. W. (1986). Phase discrimination of compound gratings: Generalized autocorrelation analysis. *Journal of the Optical Society of America A, 3*, 868–879.

Lohmann, A. W., & Wirnitzer, B. (1984). Triple correlation. *Proceedings of the IEEE, 72*, 889–901.

Resnikoff, H. L. (1989). *The illusion of reality*. New York: Springer-Verlag.

Yellott, J. I., Jr., & Iverson, G. J. (1989). Triple correlation and texture discrimination. *Investigative Ophthalmology and Visual Science, 31*, 561, (No. 4). (Abstract of a paper presented at the 1990 annual meeting of the Association for Research in Vision and Opthalmology.)

8 Exemplars, Prototypes, and Similarity Rules

Robert M. Nosofsky
Indiana University

One of the major themes in categorization research during the past two decades has involved comparisons and contrasts between prototype and exemplar models (e.g., Estes, 1986a; Hintzman, 1986; Homa & Chambliss, 1975; Medin & Schaffer, 1978; Nosofsky, 1987; Posner & Keele, 1970; Reed, 1972). According to prototype models, the observer forms an abstract summary representation of a category, usually assumed to be the central tendency of the category distribution. According to exemplar models, the observer stores the individual training exemplars of a category in memory. In both prototype and exemplar models, classification decisions are based on the similarity of an item to the underlying category representation.

A related theme has involved a contrast between independent-cue and interactive-cue models (Medin & Schaffer, 1978; Medin & Smith, 1981). As discussed by Medin and Smith, "independent-cue theories assume that the information entering into category judgments (overall similarity, distance, or validity) can be derived from an additive combination of the information from component attributes" (p. 241). By contrast, interactive-cue models reject this assumption of additivity.

In an elegant and highly influential program of research, Medin and his associates (Medin, Altom, Edelson, & Freko, 1982; Medin, Altom, & Murphy, 1984; Medin, Dewey, & Murphy, 1983; Medin & Schaffer, 1978; Medin & Schwanenflugel, 1981; Medin & Smith, 1981) systematically contrasted the predictions of a fairly general independent-cue model with those of an interactive-cue exemplar model known as the context model. This program of research, some of which is summarized later in this chapter, overwhelmingly supports the predictions of the context model over the independent-cue model. Because certain types of prototype models are special cases of the independent-cue model, Medin's research provides evidence in favor of exemplar-based category representations.

A natural question that arises, however, is whether the superiority of the context model derives from its assumption of an exemplar-based category representation or from its use of an interactive-cue combination rule. The independent-cue model formalized by Medin can be viewed as a prototype model that uses an additive similarity rule. By contrast, the context model is an exemplar model that uses an interactive (multiplicative) similarity rule. Thus, both the nature of the category representation and the form of the similarity rule differ in the two models.

One purpose of this chapter is to fill in the missing parts of the picture. Following in the spirit of Estes' (1986a) "Array Models" paper, the goal is to achieve as close a comparison between competing models as possible by testing models that differ with respect to one critical assumption. Thus, in addition to considering the multiplicative-similarity exemplar model and additive-similarity prototype model tested by Medin, I also consider an additive-similarity exemplar model and a multiplicative-similarity prototype model. The upshot of the investigation is that both the multiplicative-similarity rule and the exemplar-based category representation are critical in allowing the context model to achieve its accurate quantitative predictions of classification performance.

The present chapter does more than clarify previous theoretical investigations, however. In particular, it is shown that two current and highly influential models of classification—the simple (nonconfigural) adaptive network models proposed by Gluck and Bower (1988a), and the fuzzy logical model of perception (FLMP) investigated by Massaro (1987; Massaro & Friedman, 1990; Oden & Massaro, 1978)—are either formally identical to or are special cases of the multiplicative-similarity prototype model considered herein. Although these models use multiplicative-combination rules, it is argued that their underlying prototype representation leads to shortcomings in their predictions.

The chapter is organized as follows. The first section sets up the general theoretical framework in which the models are organized. The next section considers in some detail the nature of the multiplicative-similarity prototype model and its relations to Gluck and Bower's adaptive network models and Massaro's FLMP. The third section reports quantitative fits of the alternative models to various classification data sets published previously by Medin and his associates, and to some data sets reported by Gluck and Bower (1988a). In a nutshell, it is shown that only the multiplicative-similarity exemplar model (the context model) fares well across all data sets. The locus of the shortcomings of the alternative models is discussed. Finally, the last section briefly reviews research conducted by Nosofsky that also provides overwhelming evidence in favor of multiplicative-similarity exemplar models over multiplicative-similarity prototype models.

THEORETICAL FRAMEWORK

The formal development is limited to situations involving stimuli that vary along M binary-valued dimensions. Extensions to continuous-valued dimensions are considered in the final section of the chapter. I also assume that objects are being classified into two categories.

In formal tests, the prototype of a category is generally defined as the central tendency (or centroid) of the category training exemplars. Note that the central tendency corresponds to a <u>single point</u> in the M-dimensional space. I generalize the definition of a prototype in this chapter by allowing the prototype to correspond to <u>any</u> single point in the M-dimensional space in which the exemplars are embedded. Thus, the prototype could be the centroid of all category exemplars, a vector of modal values over all category exemplars, a vector of "ideal" values, and so forth. Although still more general definitions of "prototype" are possible (cf. Barsalou, 1989), the present space of models seems like a reasonable one to explore.

The prototypes of Categories 1 and 2 are denoted by the vectors $\mathbf{p}_1 = (p_{11}, p_{12}, ..., p_{1M})$ and $\mathbf{p}_2 = (p_{21}, p_{22}, ..., p_{2M})$, respectively. A particular target probe is denoted by the vector $\mathbf{t} = (t_1, t_2, ..., t_M)$, whereas an individual training exemplar i is denoted by $\mathbf{i} = (i_1, i_2, ..., i_M)$. Nonnegative similarity parameters $s(t_m, p_{1m})$ and $s(t_m, p_{2m})$ are defined that represent the similarity between the probe and prototypes on dimension m. Likewise, $s(t_m, i_m)$ denotes the similarity between the probe and exemplar i on dimension m.

According to the prototype models, the probability that probe $\mathbf{t}$ is classified in Category 1 is found by computing the similarity between $\mathbf{t}$ and the Category 1 prototype and then dividing by the sum of similarities between $\mathbf{t}$ and both category prototypes. In the additive-similarity prototype model, an additive-similarity rule is used to compute the similarity between the probe and each prototype, whereas in the multiplicative-similarity prototype model, a multiplicative rule is used. Thus, according to the additive-similarity prototype model, the probability that probe $\mathbf{t}$ is classified in Category 1 is given by

$$P(R_1|\mathbf{t}) = \sum_m s(t_m, p_{1m}) / \{\sum_m s(t_m, p_{1m}) + \sum_m s(t_m, p_{2m})\}, \qquad (1)$$

whereas in the multiplicative-similarity prototype model, this classification probability is given by

$$P(R_1|\mathbf{t}) = \prod_m s(t_m, p_{1m}) / \{\prod_m s(t_m, p_{1m}) + \prod_m s(t_m, p_{2m})\}. \qquad (2)$$

In the exemplar models, the probability that $\mathbf{t}$ is classified in Category 1 is found by summing the similarities of $\mathbf{t}$ to all exemplars of Category 1, and then dividing by the sum of similarities of $\mathbf{t}$ to all exemplars of both categories. In the additive-similarity exemplar model, the probability that $\mathbf{t}$ is classified in Category 1 is given by

$$P(R_1|t) = \sum_{i \in C_1} \sum_m s(t_m, i_m) / \{ \sum_{i \in C_1} \sum_m s(t_m, i_m) + \sum_{j \in C_2} \sum_m s(t_m, j_m)\}, \quad (3)$$

whereas in the multiplicative-similarity exemplar model, this classification probability is given by

$$P(R_1|t) = \sum_{i \in C_1} \prod_m s(t_m, i_m) / \{ \sum_{i \in C_1} \prod_m s(t_m, i_m) + \sum_{j \in C_2} \prod_m s(t_m, j_m)\}. \quad (4)$$

As proved in the Appendix (Proof 1), in their most general form the additive-similarity prototype and additive-similarity exemplar models (Equations 1 and 3) are formally identical, and so it will suffice hereafter to consider only the additive-similarity prototype model. Furthermore, I show in the Appendix (Proof 2) that the pure version of the independent-cue model tested by Medin and his associates is a special case of the additive-similarity prototype model.[1] The multiplicative-similarity prototype and exemplar models (Equations 2 and 4) are not formally identical. Indeed, I show later in this chapter that these models make dramatically different qualitative predictions of classification performance for certain category structures.

The multiplicative-similarity exemplar model (Equation 4) is the context model proposed by Medin and Schaffer (1978), except some of the constraints on the free parameters have not been introduced. Recall that the dimensions are binary-valued. Without loss of generality, it can be assumed that $s(t_m, i_m) = 1$ if $t_m = i_m$. Medin and Schaffer further assumed that if $t_m \neq i_m$, then $s(t_m, i_m) = s(i_m, t_m) = s_m$, where s_m is a freely estimated similarity parameter for mismatching values on dimension m. The model could be generalized to allow asymmetric similarity relations, but in the remainder of this chapter I adopt the assumption of symmetry when testing the context model.

ON THE MULTIPLICATIVE-SIMILARITY PROTOTYPE MODEL

This section demonstrates formal relations between the multiplicative-similarity prototype (MSP) model and some current and highly influential models of classification performance. Because the material is somewhat technical in nature, the reader may wish to skip directly to the next section for a summary of the key points.

For purposes of demonstrating theoretical relations among extant models, it is convenient to consider a general version of the MSP model that allows asymmetric similarity relations, as well as considering a symmetric-similarity version.

---

[1] To equate the number of parameters used by the context model and the independent-cue model, Medin and Smith (1981) added a holistic exemplar-coding parameter to the independent-cue model. This same parameter could also be added to the additive-similarity prototype model.

FIG. 8.1. Schematic illustration of the symmetric-similarity condition.

T_{m1} P_{1m} P_{2m} T_{m2}

Dividing numerator and denominator of the MSP model (Equation 2) by Π $s(t_m, p_{1m})$ yields:

$$P(R_1|t) = \{1 + \prod_m [s(t_m, p_{2m})/s(t_m, p_{1m})]\}^{-1}$$

$$= \{1 + \prod_m r(t_m)\}^{-1}, \tag{5}$$

where $r(t_m)$ denotes the ratio of similarities on dimension m $[s(t_m,p_{2m})/s(t_m,p_{1m})]$. Because the dimensions are binary-valued, each t_m can take on one of two possible values (denote them t_{m1} and t_{m2}), so there are 2M similarity ratios. I refer to the general version of the model in which all ratios are free to vary as the general MSP model.

An important special case of the general MSP model arises when there is a symmetry in similarity relations between values of the target probes and the respective prototypes; see Fig. 8.1. In particular, a symmetry condition is defined to hold on dimension m whenever $s(t_{m1},p_{1m}) = s(t_{m2},p_{2m})$ and $s(t_{m2},p_{1m}) = s(t_{m1},p_{2m})$, in which case $r(t_{m1}) = 1/r(t_{m2})$. I refer to Equation 2 as the symmetric MSP model when the symmetry condition holds for all M dimensions. The symmetric MSP model requires estimation of M free parameters, one similarity-ratio parameter for each dimension. Medin et al. (1984) incorporated the symmetric MSP model in some tests of a mixture model of classification. A baseline version of the symmetric MSP model with constant similarity-ratio parameters across dimensions was also discussed and analyzed by Estes (1986a).

Although the general MSP model has 2M similarity ratios, it actually has only $M + 1$ free parameters. In particular, as shown in the appendix (Proof 3), the general MSP model can always be rewritten as a symmetric MSP model with an additional response-bias parameter:

$$P(R_1|t) = \{1 + b \prod_m r'(t_m)\}^{-1}, \tag{6}$$

where $r'(t_{m1}) = 1/r'(t_{m2})$ for all m, and the response-bias parameter (b) is nonnegative. For values of $b > 1$ there is a bias toward responding Category 2, whereas for $b < 1$ there is a bias toward Category 1.

Relation to Gluck and Bower's (1988a) Adaptive Network Models.[2]   The one-layered network models proposed by Gluck and Bower (1988a) are illustrated for a four-dimensional case in Fig. 8.2. Presentation of a probe gives rise

---

[2]Although the work was conducted independently, the formal development presented in this section overlaps in part with previous work presented by Golden and Rumelhart (1989) and Massaro and Friedman (1990).

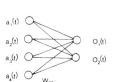

A. Single - Node Model

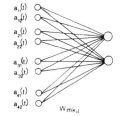

B. Double - Node Model

FIG. 8.2. Schematic illustration of one-layered (nonconfigural) adaptive network models proposed by Gluck and Bower (1988a). Panel A: single-node model, Panel B: double-node model.

to a pattern of activations over the input nodes. These activations are multiplied by weights on the connecting links and then summed to determine outputs. For the single-node model (Fig. 8.2A), the output to category node j is given by

$$O_j(t) = \sum_m a_m(t)w_{mj}, \tag{7}$$

where $a_m(t)$ denotes the activation received by input node m upon presentation of pattern t, and w_{mj} denotes the weight from input node m to category node j. In the models tested by Gluck and Bower and also by Estes, Campbell, Hatsopoulos, and Hurwitz (1989), the weights were updated trial to trial using the delta rule (Rumelhart, Hinton, & Williams, 1986; Widrow & Hoff, 1960). In the present chapter, instead of viewing the networks as learning models, I focus on their general architecture and consider their ability to predict classification transfer data in their most general form. In other words, the weights are allowed to be free parameters rather than being constrained by a particular learning rule. If the general architecture of the models is shown to have shortcomings, then one knows in advance that the more restricted models that posit particular learning rules will have the same shortcomings.

Following Gluck and Bower (1988a) and Estes et al. (1989), the outputs defined in Equation 7 are exponentiated and entered into a choice rule to predict classification probabilities:

$$P(R_1|t) = \exp[cO_1(t)]/\{\exp[cO_1(t)] + \exp[cO_2(t)]\}, \tag{8}$$

where c is a scale parameter. Dividing numerator and denominator of Equation 8 by $\exp[cO_1(t)]$ yields

$$P(R_1|t) = \{1 + \exp[cO_2(t) - cO_1(t)]\}^{-1}$$
$$= \{1 + \exp[c(\Sigma a_m(t)w_{m2} - \Sigma a_m(t)w_{m1})]\}^{-1}$$
$$= \{1 + \exp[c\Sigma a_m(t)(w_{m2} - w_{m1})]\}^{-1}$$
$$= \{1 + \exp[\Sigma a_m(t)v_m]\}^{-1}$$
$$= \{1 + \prod_m \exp[a_m(t)v_m]\}^{-1}, \tag{9}$$

where $v_m = c(w_{m2} - w_{m1})$.

Comparing Equation 9 to Equation 5, it is seen that the single-node model is formally identical to the general MSP model, with $r(t_m) = \exp[a_m(t)v_m]$. Two special cases of the single-node model are of interest. In the first, the activations to the input nodes $[a_m(t)]$ are either 0 or 1 (as assumed by Gluck and Bower, 1988a), in which case each $r(t_m)$ is equal either to one or a free parameter $r_m = \exp(v_m)$. In the second, each $a_m(t)$ is equal either to -1 or 1 (e.g., Markman, 1989), in which case each $r(t_m)$ is equal either to $r_m = \exp(v_m)$ or $(1/r_m) = \exp(-v_m)$. This latter special case model is formally identical to the symmetric MSP model. Each of these special-case network models has M free parameters (one value of r_m for each dimension).

For the double-node network model (Fig. 8.2B), two nodes code the values on each dimension. For the version tested by Gluck and Bower (1988a) and Estes et al. (1989), the presence of value 1 on dimension m leads to activation of node 1 on dimension m and no activation of node 2 on dimension m, and vice versa for the presence of value 2. Alternatively, as suggested by Markman (1989), one could use a $(-1,1)$ activation scheme. In general, the output to category node j is given by

$$O_j(t) = \sum_m [a_{m1}(t)w_{m1,j} + a_{m2}(t)w_{m2,j}], \tag{10}$$

where $a_{mk}(t)$ is the activation received by input node k on dimension m when presented with probe t, and $w_{mk,j}$ is the weight from node k on dimension m to category node j.

As was the case for the single-node model, the probability that probe t is classified into Category 1 is given by

$$P(R_1|t) = \{1 + \exp[cO_2(t) - cO_1(t)]\}^{-1}$$
$$= \{1 + \exp[c(\Sigma a_{m1}(t)w_{m1,2} + a_{m2}(t)w_{m2,2}) - c(\Sigma a_{m1}(t)w_{m1,1} + a_{m2}(t)w_{m2,1})]\}^{-1}$$
$$= \{1 + \exp[c(\Sigma a_{m1}(t)(w_{m1,2} - w_{m1,1}) + a_{m2}(t)(w_{m2,2} - w_{m2,1}))]\}^{-1}$$
$$= \{1 + \exp[\Pi a_{m1}(t)v_{m1} + a_{m2}(t)v_{m2}]\}^{-1}$$
$$= \{1 + \prod_m \exp[a_{m1}(t)v_{m1} + a_{m2}(t)v_{m2}]\}^{-1}, \tag{11}$$

where $v_{mk} = c(w_{mk,2} - w_{mk,1})$.

TABLE 8.1
Relations Between MSP Models and Adaptive Network Models

General MSP model	double-node model with (0, 1) activation codings
Symmetric MSP model	single and double-node models with (-1, 1) activation codings
MSP model with $r(tm) = r_m$ if tm = 1 $r(tm) = 1$ if tm = 0	single-node model with (0, 1) activation codings

Comparing Equations 11 and 5, we see that the double-node model with $(0,1)$ activation codings is formally identical to the general MSP model. If $a_{m1}(t) = 1$ and $a_{m2}(t) = 0$, then $r(t_m) = \exp(v_{m1})$; whereas if $a_{m1}(t) = 0$ and $a_{m2}(t) = 1$, then $r(t_m) = \exp(v_{m2})$. Furthermore, the double-node model with $(-1,1)$ activation codings is formally identical to the symmetric MSP model. Either $r(t_m) = \exp(v_{m1} - v_{m2})$ or $r(t_m) = \exp(v_{m2} - v_{m1}) = 1/\exp(v_{m1} - v_{m2})$.

A summary of the relations between the network models and the multiplicative-similarity prototype model is provided in Table 8.1.

Relation to Massaro's Fuzzy Logical Model of Perception (FLMP). As described by Massaro (1987, pp. 20–22), in the FLMP each category is represented by a prototype of ideal values along M dimensions. Classification is determined by the relative degree to which the features of a stimulus match the ideal values of the respective prototypes. Following Massaro, but using a different notation, let $f_{tm}(0 \leq f_{tm} \leq 1)$ denote the (relative) degree to which probe **t** matches prototype 1 on dimension m. Then in the FLMP, the relative degree to which **t** matches prototype 2 on dimension m would be given by the additive complement $1 - f_{tm}$.[3] The probability that **t** is classified into Category 1 is then given by

$$P(R_1|t) = \prod_m f_{tm} / \{\prod_m f_{tm} + \prod_m (1 - f_{tm})\}$$

$$= \{1 + \prod_m [(1 - f_{tm})/f_{tm}]\}^{-1}. \tag{12}$$

Comparison with Equation 5 reveals that when applied to stimuli varying on binary-valued dimensions, the FLMP is formally identical to the general MSP

[3]Because Massaro sets $f_{tm}(2) = 1 - f_{tm}(1)$, where $f_{tm}(j)$ is the degree of match of probe **t** to prototype j on dimension m, the degrees of match should be interpreted as relative rather than absolute. For example, suppose that Prototypes 1 and 2 have identical values on dimension m, and that **t** matches Prototype 1 perfectly on that dimension. If the interpretation of the parameter $f_{tm}(1)$ is one of absolute degree of match, then $f_{tm}(1) = 1$, which implies $f_{tm}(2) = 1 - f_{tm}(1) = 0$, which is a contradiction because Prototypes 1 and 2 are identical on dimension m. It is also unclear how Massaro would generalize the scheme of using additive complements so as to apply the FLMP to more than two categories.

model, with $r(t_m) = (1 - f_{tm})/f_{tm}$. In applying the FLMP to some of Medin's data, Massaro (1987, pp. 251–259) apparently introduced the further constraint that $f_{tm1} = 1 - f_{tm2}$, where f_{tm1} and f_{tm2} denote the (relative) degrees of match of the two possible values on dimension m (t_{m1} and t_{m2}) to prototype 1 on dimension m. This constraint implies that $r(t_{m1}) = 1/r(t_{m2})$, which is the constraint that defines the symmetric MSP model.

COMPARING THE ALTERNATIVE MODELS

To summarize the development thus far, I started by setting up four main models: the additive-similarity prototype and exemplar models, and the multiplicative-similarity prototype and exemplar models. The additive-similarity prototype and exemplar models are formally identical, and are generalizations of the (pure) independent-cue model tested by Medin and Smith (1981). I also noted that a general version of the multiplicative-similarity prototype (MSP) model that allows asymmetric similarity relations can always be rewritten as a symmetric MSP model with a response-bias parameter. Furthermore, when the weight parameters are not constrained by a learning rule, the one-layered network models tested by Gluck and Bower (1988a), as well as Massaro's (1987) fuzzy logical model of perception, are either formally identical to or are special cases of the general MSP model.

The main aim of this section is to compare the alternative models on their ability to predict sets of classification data published previously by Medin and his associates, as well as some data sets published by Gluck and Bower (1988a). First, however, it is useful to review an important constraint on classification implied by the models: The prototype models are able to predict above-chance performance on all of the category training exemplars if and only if the categories are linearly separable, <u>regardless of whether a multiplicative-similarity rule or an additive-similarity rule is used.</u> By contrast, the multiplicative-similarity exemplar model is able to predict above-chance performance on all exemplars, for both linearly separable and nonlinearly separable categories.

A proof that linear-separability is a necessary requirement for accurate classification by the multiplicative-similarity prototype model is provided in the appendix (Proof 4). To take an extreme example, consider the exclusive-or category structure shown in Table 8.2. Imagine that the dimensions are continuous-valued so that an averaging process seems sensible. Note that the central tendencies of Categories A and B are identical, namely (½,½). Obviously, if the category representations cannot be discriminated, it makes little difference what similarity rule is used—performance would be at chance. By way of contrast, note that if the similarity parameters for mismatching values on each dimension m (s_m) were set at zero, the context model (Equation 4) would predict perfect performance on the exclusive-or problem.

TABLE 8.2
Exclusive-OR Category Structure

	Category A			Category B	
	Dimension Values			Dimension Values	
	1	2		1	2
Stimulus			Stimulus		
1	1	0	3	1	1
2	0	1	4	0	0
Central Tendency	1/2	1/2		1/2	1/2

Now, conceivably, the classification strategy that subjects naturally adopt is to form category prototypes, and they may resort to the storage of exemplar information only if the prototype strategy fails to yield adequate performance. Thus, although exemplar models have been shown in previous work to yield accurate quantitative predictions of classification even for nonlinearly separable categories (e.g., Estes, 1986b; Medin et al., 1982; Nosofsky, 1986, 1987, 1989; Shepard & Chang, 1963), this section considers only linearly-separable cases, where presumably all the competing models are applicable.

Table 8.3 reports quantitative fits of the alternative models to 13 previously published sets of classification transfer data. All data sets were collected in "standard" classification learning paradigms in which subjects learned the categories trial by trial via induction over individual training exemplars. In all cases the categories were linearly separable, and the stimuli were composed of four binary-valued dimensions.[4] Therefore, the general MSP has five free parameters, whereas the symmetric MSP model, the single-node model with (0,1) codings, and the context model all have four free parameters. For the single-node model with (0,1) codings, fits are provided only for Gluck and Bower's (1988a) Experiments 1 and 2. In these experiments, each dimension coded the presence or absence of an individual feature, the situation in which the single-node model with (0,1) codings is plausibly applied.

The model fits are reported in terms of sum of squared deviations between predicted and observed Category 1 response probabilities. The results of the reanalyses are clear-cut. Not surprisingly, despite its abundance of free parameters, the additive-similarity prototype model performs poorly relative to the

---

[4]Gluck and Bower's (1988a) experiments had the additional feature that the categories were defined over independent probability distributions of the values on the four dimensions.

TABLE 8.3
Sum of Squared Deviations Between Predicted and Observed Category 1 Response
Probabilities for the Alternative Models

	Model				
Data Set	Additive Prototype	General MSP	Symmetric MSP	Context	(0-1) Coding Single-Node MSP
Medin and Schaffer (1978)					
Exp. 2	.260	.204	.205	.060	
Exp. 3	.154	.121	.124	.031	
Exp. 4	.099	.197	.242	.078	
Medin and Smith (1981)					
Proto. Instruct.	.032	.062	.081	.029	
Rule Instruct.	.139	.100	.101	.049	
Medin, Dewey, and Murphy (1983)					
last-name infinite	.073	.034	.043	.043	
last-name only	.034	.042	.046	.053	
Medin, Altom, and Murphy (1984)					
Exp. 1 (examples only)	.108	.111	.115	.052	
Exp. 2 (examples only,geometric)	.117	.103	.107	.041	
Exp. 3 (examples only, verbal)	.205	.209	.225	.072	
Gluck and Bower (1988a)					
Exp. 1	.009	.018	.703	.020	.108
Exp. 2	.102	.016	.305	.019	.051
Exp. 3	.014	.010	.841	.030	.175

context model. More important, in only one case does the symmetric MSP model provide a better fit than does the context model, and the advantage here is negligible (.046 vs. .053 in the last-name infinite condition of Medin et al., 1983). By contrast, there are numerous cases in which the context model yields a far better fit than does the symmetric MSP model. And in the three Gluck and Bower (1988a) experiments, the context model outperforms both the symmetric MSP model and the single-node model with (0,1) codings. Finally, even the general five-parameter MSP model, which is allowed a response bias, rarely outperforms the four-parameter context model, and in most cases it performs far worse.

To gain some insight into the reason for the superiority of the context model, it

TABLE 8.4
Diagnostic Category Structure Tested by Medin in Various Experimental
Conditions Reported in Table 8.3
(Training Exemplars Only)

	Category A					Category B			
	Dimension Values					Dimension Values			
Stimulus	1	2	3	4	Stimulus	1	2	3	4
1	1	1	1	0	6	1	1	0	0
2	1	0	1	0	7	0	1	1	0
3	1	0	1	1	8	0	0	0	1
4	1	1	0	1	9	0	0	0	0
5	0	1	1	1					
Modal Proto- type	1	1	1	1		0	?	0	0

is useful to review one of the designs that Medin created and implemented in various experiments to contrast the predictions of the independent-cue models and the context model. The category structure is shown in Table 8.4 in terms of the abstract logical codings that defined the stimuli. The training exemplars of Category A tend to have (logical) value 1 on each of their dimensions, whereas the training exemplars of Category B tend to have value 0. The modal prototypes of Categories A and B are 1111 and 0?00, respectively.

A critical qualitative distinction between the predictions of the context and prototype models involves classification accuracy for Stimuli 1 and 2 of Category A. As explained previously by Medin and Schaffer (1978), Stimulus 1 is at least as similar to the modal prototype of Category A as is Stimulus 2, no matter how the dimensions are weighted. More pertinent to the theme of this chapter, Stimulus 1 is more similar to the modal prototype of Category A than is Stimulus 2, regardless of whether similarity is computed using an additive or a multiplicative rule. Thus, the natural prediction of the prototype models is that Stimulus 1 will be classified into Category A with higher probability than Stimulus 2.

To back up this assertion, I conducted computer simulations involving the double-node version of Gluck and Bower's (1988a) one-layered adaptive network model. Training patterns from Categories A and B were randomly selected and presented to the network, and feedback was presented on each trial. The weights in the network were updated trial by trial using the delta rule. The learning rate was set at $\beta = .02$ and the scale parameter in the choice function (Equation 8) was set at $c = 2.00$. (These values are similar to the ones used by Gluck and Bower, 1988a.) After 240 training trials, the network was used to generate predictions of Category A response probabilities for each of the patterns

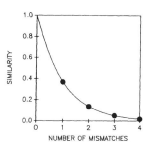

FIG. 8.3. Form of the similarity gradient predicted by the multiplicative rule. Similarity is plotted as a function of number of mismatching values on the component dimensions. (The similarity parameters s_m are assumed to be constant across dimensions, with $0 < s_m < 1$.) The solid curve is an exponential decay function (see section titled "Extensions to Continuous-Dimension Domains").

shown in Table 8.4. The critical result was that, after averaging over the results of 1000 simulations, the network did indeed predict a higher probability of a Category A response for Stimulus 1 (.79) than for Stimulus 2 (.76).

The opposite prediction arises from the context model, because of its joint assumption of an exemplar-based category representation and a multiplicative-similarity rule. Figure 8.3 illustrates the overall similarity between two patterns as a function of the number of mismatching values across the four dimensions. For simplicity, it is assumed that the similarity parameter is constant across dimensions. A critical aspect of the similarity gradient predicted by the multi-plicative rule is that for each additional mismatch, similarity decreases at a decreasing rate. This property of the similarity gradient implies that exemplars that mismatch a probe on only one dimension carry the bulk of the influence in determing the summed similarity of a probe to all exemplars of the category.[5] For example, the summed similarity of a probe to an exemplar it mismatches on one dimension and to an exemplar it mismatches on three dimensions is greater than the probe's summed similarity to two exemplars it mismatches on two dimensions.

Refer to pairs of items with only one mismatching value as neighbors. With regard to the category structure in Table 8.4, note that Stimulus 1 has only one neighbor in Category A, yet two neighbors in Category B. On the other hand, Stimulus 2 has two neighbors in Category A, and no neighbors in Category B. Thus, because of its exemplar-based category representation and its assumption of a multiplicative-similarity rule, the context model predicts that Stimulus 2 will be classified into Category A with higher probability than Stimulus 1, in direct contrast to the predictions of the prototype models.

In virtually all of the standard classification learning experiments reported by Medin, the qualitative prediction of the context model regarding performance on Stimuli 1 and 2 was supported. Thus, it is not surprising that the prototype

---

[5]This assertion excludes, of course, the stored exemplars that perfectly match the probes.

models tend to fare worse than the context model in terms of their quantitative fits, regardless of the form of the similarity rule that is assumed.[6]

EXTENSIONS TO CONTINUOUS-DIMENSION DOMAINS

The model comparisons considered in the previous section were limited to domains involving binary-valued stimulus dimensions. Nosofsky (1984, 1986) noted that the multiplicative-similarity rule proposed by Medin and Schaffer (1978) can be viewed as a special case of a multidimensional scaling (MDS) approach to modeling similarity. For example, the multiplicative-similarity rule arises if distance in psychological space is computed using a city-block metric (e.g., Garner, 1974; Shepard, 1964), and similarity is an exponential decay function of distance in the space (e.g., Shepard, 1958, 1987). (The exponential similarity gradient is the one illustrated in Fig. 8.3.) This MDS-based interpretation of the multiplicative-similarity rule led Nosofsky (1984, 1986) to propose the generalized context model (GCM), wherein exemplars are represented as points in a multidimensional psychological space, and similarity is assumed to be a decreasing function of distance in the space.

Numerous tests of the GCM have involved systematic comparisons between exemplar-based and prototype-based category representations in continuous-dimension stimulus domains, including geometric forms varying continuously in size and orientation, colors varying in brightness and saturation, schematic faces varying along continuous attributes such as eye separation, nose length, and mouth height, and random dot patterns with varying degrees of distortion from a prototype pattern (e.g., Nosofsky, 1986, 1987, 1988, 1991; Shin, 1990). The general approach in these comparisons has involved first deriving MDS solutions for the sets of exemplars on the basis of errors in identification confusion experiments or on the basis of direct similarity judgments. In predicting performance in subsequent categorization tasks, similarities among the items are then computed from the derived MDS solutions. A critical aspect of all these comparisons is that the exemplar and prototype models use the same free parameters and the same

---

[6]Gluck and Bower (1988b) proposed an augmented network model in which the input nodes code not only individual values on each dimension, but also combinations of values. This augmented model does not have a prototype-based category representation as it is defined here. Also, unless the weights are constrained by a learning rule (such as the delta rule used by Gluck & Bower, 1988b), this model has too many free parameters to be included in the present quantitative comparisons. To date, quantitative tests of the augmented network model with configural cues have not been conducted.

In a similar vein, Massaro (1987, pp. 254–258) proposed and illustrated that combinations of dimensions could sometimes be redefined to allow prototype models to accurately predict performance. Whether accurate fits to most of Medin's and Nosofsky's data sets could indeed be achieved using this approach remains to be demonstrated. More important, however, some independent basis is needed for predicting when and how such redefinition of dimensions takes place.

TABLE 8.5
Comparison of Model Fits for the Continuous-Dimension Exemplar and Prototype Models

	Model	
Data Set	*Exemplar*	*Prototype*
Nosofsky (1987)		
Saturation (A)	42.5	43.3
Saturation (B)	69.0	130.8
Cross-Cross	43.3	59.4
Diagonal	65.5	139.5
Brightness	54.6	48.6
Pink-Brown	57.8	154.1
Nosofsky (1988)		
Exp. 1	193.9	299.8
Exp. 2	199.7	277.5
Nosofsky (1991)		
Exp. 1A	136.0	191.8
Exp. 1B	129.2	181.0
Exp. 2A	219.1	303.7
Exp. 2B	194.8	262.8
Shin (1990)		
Exp. 1	225.7	527.4
Exp. 2, Immediate Condition	169.7	226.2
Exp. 2, Delayed Condition	181.3	203.2
Exp. 3, Size - 3		
Equal-Frequency Condition	63.4	82.7
Exp. 3, Size - 3		
Unqual Frequency Condition	62.2	83.4
Exp. 3, Size - 10		
Equal Frequency Condition	84.2	119.6
Exp. 3, Size - 10		
Unequal-Frequency Condition	75.0	138.0

Note. The criterion of fit was to minimize $-\ln L$, where $\ln L$ is the (natural) log-likelihood of the data set. Multiple-prototype models were fitted to the criss-cross and diagonal conditions of Nosofsky (1987), because in these conditions the categories were nonlinearly separable.

rule for computing similarity in the multidimensional psychological space—only the nature of the category representation differs. A summary of model fits across 19 conditions involving a variety of category structures, experimental conditions, and types of stimuli is provided in Table 8.5. Although a detailed discussion of the model-fitting analyses goes beyond the scope of this chapter, it is easily seen that the evidence is overwhelmingly in favor of exemplar-based category representations compared to prototype-based category representations, with the nature of the similarity rule held constant.

A FINAL COMMENT

It is fitting that the present model comparisons are reported in this volume. Almost a decade ago, at a time when I worked as his research assistant, William K. Estes noticed the relation between Medin and Schaffer's (1978) context model and the independent-cue model. In the spirit of developing closer model comparisons, he suggested the multiplicative-similarity prototype model, and we conducted many of the quantitative comparisons reported herein. To my knowledge, these quantitative comparisons have never been published. With the advent of the highly influential multiplicative-similarity prototype models proposed by Gluck and Bower (1988a) and Massaro (1987; Massaro & Friedman, 1990), they take on renewed relevance. I also hope that the present chapter adds perspective to the outcomes of the extensive quantitative comparisons between prototype and exemplar models that have been conducted during the past decade, and that this perspective will advance progress in psychological theorizing about the nature of category representations.

ACKNOWLEDGMENTS

Preparation of this chapter was supported by Grants BNS 87-19938 from the National Science Foundation and PHS R01 MH48494-01 from the National Institute of Mental Health to Indiana University. My thanks to William Estes, Mark Gluck, Dominic Massaro, and Douglas Medin for their helpful comments on an earlier version of this chapter.

APPENDIX

Proof 1. The additive-similarity exemplar model (Equation 3) can be rewritten as

$$P(R_1|t) = \sum_m \sum_{i\in C_1} s(t_m, i_m) / \{\sum_m \sum_{i\in C_1} s(t_m, i_m) + \sum_m \sum_{j\in C_2} s(t_m, j_m)\}. \quad (A1)$$

Letting $s(t_m, p_{1m}) = \sum_{i\in C_1} s(t_m, i_m)$ establishes the formal identity between the additive-similarity prototype and exemplar models.

Proof 2. According to Medin and Smith's (1981) independent-cue model,

$$P(R_1|t) = \sum_m \delta_m(t)w_m, \quad (A2)$$

where $0 \le w_m \le 1$ and $\sum w_m = 1$, and the $\delta_m(t)$ are indicator variables equal to one if **t** has (logical) value one on dimension m, and equal to zero otherwise. The additive-similarity prototype model can be written in expanded form as

$$P(R_1|t) = \{\sum_m \delta_m(t)s(t_{m1}, p_{1m}) + [1 - \delta_m(t)]s(t_{m2}, p_{1m})\} /$$

$$\{ \sum_m \delta_m(t) s(t_{m1}, p_{1m}) + [1 - \delta_m(t)] s(t_{m2}, p_{1m})$$

$$+ \sum_m \delta_m(t) s(t_{m1}, p_{2m}) + [1 - \delta_m(t)] s(t_{m2}, p_{2m}) \}. \quad \text{(A3)}$$

Set $s(t_{m2}, p_{1m}) = s(t_{m1}, p_{2m}) = 0$ and $s(t_{m1}, p_{1m}) = s(t_{m2}, p_{2m}) = w_m$ (again, where $0 \le w_m \le 1$ and $\Sigma w_m = 1$). Then

$$P(R_1|t) = \sum_m \delta_m(t) w_m / \sum_m w_m = \sum_m \delta_m(t) w_m, \quad \text{(A4)}$$

which is the independent-cue model equation.

Proof 3. According to the general MSP model,

$$P(R_1|t) = \{ 1 + \prod_m r(t_{m1})^{\delta_m(t)} r(t_{m2})^{1-\delta_m(t)} \}^{-1}, \quad \text{(A5)}$$

where $\delta_m(t)$ is the indicator variable defined in Proof 2. According to the biased, symmetric MSP model,

$$P(R_1|t) = \{ 1 + b \prod_m r'(t_{m1})^{\delta_m(t)} [1/r'(t_{m1})]^{1-\delta_m(t)} \}^{-1}$$

$$= \{ 1 + b \prod_m r'(t_{m1})^{2\delta_m(t)} r'(t_{m1})^{-1} \}^{-1}. \quad \text{(A6)}$$

The formal identity between the models is established by showing that there exist b and $r'(t_{m1})$, $m = 1,M$, such that for all t,

$$b \prod_m r'(t_{m1})^{2\delta_m(t)} r'(t_{m1})^{-1} = \prod_m r(t_{m1})^{\delta_m(t)} r(t_{m2})^{1-\delta_m(t)}. \quad \text{(A7)}$$

Rewriting the righthand side of (A7), we need to show

$$b \prod_m r'(t_{m1})^{2\delta_m(t)} r'(t_{m1})^{-1} = \prod_m [r(t_{m1})/r(t_{m2})]^{\delta_m(t)} r(t_{m2}). \quad \text{(A8)}$$

Set $r'(t_{m1}) = [r(t_{m1})/r(t_{m2})]^{1/2}$. Then substituting into (A8) gives

$$b \prod_m [r(t_{m1})/r(t_{m2})]^{\delta_m(t)} [r(t_{m1})/r(t_{m2})]^{-1/2} = \prod_m [r(t_{m1})/r(t_{m2})]^{\delta_m(t)} r(t_{m2}), \quad \text{(A9)}$$

and solving for b gives

$$b = \prod_m r(t_{m2}) / \{ \prod_m [r(t_{m1})/r(t_{m2})]^{-1/2} \}. \quad \text{(A10)}$$

Proof 4. By "accurate" classification of a set of patterns by the general MSP model (the biased, symmetric MSP), we mean that there exists a set of (positive) similarity-ratio parameters $r(t_{m1})$, $m = 1,M$, and a (positive) bias parameter b, such that for all $t \in C_1$,

$$P(R_1|t) = \{ 1 + b \prod_m r(t_m) \}^{-1} > 1/2, \quad \text{(A11)}$$

and for all $t \epsilon C_2$,

$$P(R_1|t) = \{ 1 + b \prod_m r(t_m) \}^{-1} < 1/2. \quad \text{(A12)}$$

Note that each $r(t_m)$ takes on two possible values, $r(t_{m1})$ or $r(t_{m2}) = 1/r(t_{m1})$. Rewriting (All), there exist positive $r(t_m)$ and b such that for all $t \epsilon C_1$,

$$1 + b \prod_m r(t_m) < 2, \Rightarrow \sum_m \log r(t_m) < \log(1/b). \quad \text{(A13)}$$

Now each t_m has two possible (logical) values, denoted t_{m1} and t_{m2}. Without loss of generality, in the real-valued multidimensional space in which the patterns reside, locate t_{m1} at $x_m > 0$ and t_{m2} at $-x_m < 0$. Then note that there exist real-valued w_m such that $\log[r(t_{m1})] = w_m x_m$, and therefore, $\log[r(t_{m2})] = -\log[r(t_{m1})] = -w_m x_m$. Thus, (A13) can be rewritten as stating that for all $t \in C_1$, there exist w_m and b such that

$$\sum_m w_m t_m < \log(1/b). \tag{A14}$$

(Note that each t_m is a placeholder for either x_m or $-x_m$.) An analogous line of argument starting from (A12) leads to the assertion that for all $t \epsilon C_2$,

$$\sum_m w_m t_m > \log(1/b). \tag{A15}$$

Taken together, (A14) and (A15) define the condition of linear separability that we wished to prove.

REFERENCES

Barsalou, L. W. (1989). On the indistinguishability of exemplar memory and abstraction in category representation. To appear in T. K. Srull & R. S. Wyer (Eds.), *Advances in Social Cognition*, Vol. 3. Hillsdale, NJ: Erlbaum.

Estes, W. K. (1986a). Array models for category learning. *Cognitive Psychology, 18,* 500–549.

Estes, W. K. (1986b). Memory storage and retrieval processes in category learning. *Journal of Experimental Psychology: General, 115,* 155–174.

Estes, W. K., Campbell, J. A., Hatsopoulos, N., & Hurwitz, J. B. (1989). Base-rate effects in category learning: A comparison of parallel network and memory storage-retrieval models. *Journal of Experimental Psychology: Learning, Memory, and Cognition, 15,* 556–571.

Garner, W. R. (1974). *The processing of information and structure.* New York: Wiley.

Gluck, M. A., & Bower, G. H. (1988a). From conditioning to category learning: An adaptive network model. *Journal of Experimental Psychology: General, 117,* 227–247.

Gluck, M. A., & Bower, G. H. (1988b). Evaluating an adaptive network model of human learning. *Journal of Memory and Language, 27,* 166–195.

Golden, R. M., & Rumelhart, D. E. (1989). *A relationship between the context model of classification and a connectionist learning rule.* Unpublished manuscript.

Hintzman, D. L. (1986). "Schema abstraction" in a multiple-trace memory model. *Psychological Review, 93,* 411–428.

Homa, D., & Chambliss, D. (1975). The relative contributions of common and distinctive information on the abstraction from ill-defined categories. *Journal of Experimental Psychology: Human Learning and Memory, 1,* 351–359.

Markman, A. B. (1989). LMS rules and the inverse base-rate effect: Comment on Gluck and Bower (1988). *Journal of Experimental Psychology: General, 118,* 417–421.

Massaro, D. W. (1987). *Speech perception by ear and eye: A paradigm for psychological inquiry.* Hillsdale, NJ: Lawrence Erlbaum Associates.

Massaro, D. W., & Friedman, D. (1990). Models of integration given multiple sources of information. *Psychological Review, 97,* 225–252.

Medin, D. L., Altom, M. W., Edelson, S. M., & Freko, D. (1982). Correlated symptoms and simulated medical classification. *Journal of Experimental Psychology: Learning, Memory, and Cognition, 8,* 37–50.

Medin, D. L., Altom, M. W., & Murphy, T. D. (1984). Given versus induced category representations: Use of prototype and exemplar information in classification. *Journal of Experimental Psychology: Learning, Memory, and Cognition, 3,* 333–352.

Medin, D. L., Dewey, G. I., & Murphy, T. D. (1983). Relationships between item and category learning: Evidence that abstraction is not automatic. *Journal of Experimental Psychology: Learning, Memory, and Cognition, 9,* 607–625.

Medin, D. L., & Schaffer, M. M. (1978). Context theory of classification learning. *Psychological Review, 85,* 207–238.

Medin, D. L., & Schwanenflugel, P. J. (1981). Linear separability in classification learning. *Journal of Experimental Psychology: Human Learning and Memory, 7,* 355–368.

Medin, D. L., & Smith, E. E. (1981). Strategies and classification learning. *Journal of Experimental Psychology: Human Learning and Memory, 7,* 241–253.

Nosofsky, R. M. (1984). Choice, similarity, and the context theory of classification. *Journal of Experimental Psychology: Learning, Memory and Cognition, 10,* 104–114.

Nosofsky, R. M. (1986). Attention, similarity, and the identification-categorization relationship. *Journal of Experimental Psychology: General, 115,* 39–57.

Nosofsky, R. M. (1987). Attention and learning processes in the identification and categorization of integral stimuli. *Journal of Experimental Psychology: Learning, Memory, & Cognition, 13,* 87–109.

Nosofsky, R. M. (1988). Similarity, frequency, and category representations. *Journal of Experimental Psychology: Learning, Memory and Cognition, 14,* 54–65.

Nosofsky, R. M. (1989). Further tests of an exemplar-similarity approach to relating identification and categorization. *Perception & Psychophysics, 45,* 279–290.

Nosofsky, R. M. (1991). Tests of an exemplar model for relating perceptual classification and recognition memory. *Journal of Experimental Psychology: Human Perception and Performance, 17,* 3–27.

Oden, G. C., & Massaro, D. W. (1978). Integration of featural information in speech perception. *Psychological Review, 85,* 172–191.

Posner, M. I., & Keele, S. W. (1970). Retention of abstract ideas. *Journal of Experimental Psychology, 83,* 304–308.

Reed, S. K. (1972). Pattern recognition and categorization. *Cognitive Psychology, 3,* 382–407.

Rumelhart, D. E., Hinton, G. E., & Williams, R. J. (1986). Learning internal representations by error propagation. In D. E. Rumelhart & J. L. McClelland (Eds.), *Parallel distributed processing: Explorations in the microstructure of cognition: Vol. 1. Foundations* (pp. 318–362). Cambridge, MA: Bradford Books/MIT Press.

Shepard, R. N. (1958). Stimulus and response generalization: Tests of a model relating generalization to distance in psychological space. *Journal of Experimental Psychology, 55,* 509–523.

Shepard, R. N. (1964). Attention and the metric structure of the stimulus space. *Journal of Mathematical Psychology, 1,* 54–87.

Shepard, R. N. (1987). Toward a universal law of generalization for psychological science. *Science, 237,* 1317–1323.

Shepard, R. N., & Chang, J. J. (1963). Stimulus generalization in the learning of classifications. *Journal of Experimental Psychology, 65,* 94–102.

Shin, H. J. (1990). *Similarity-scaling studies of "dot patterns" classification and recognition.* Unpublished doctoral dissertation, Indiana University, Bloomington, Indiana.

Widrow, G., & Hoff, M. E. (1960). Adaptive switching circuits. *Institute of Radio Engineers, Western Electronic Show and Convention, Convention Record, 4,* 96–194.

9

Stimulus Sampling and Distributed Representations in Adaptive Network Theories of Learning

Mark A. Gluck
Rutgers University
Stanford University

Current adaptive network, or "connectionist," theories of human learning are reminiscent of statistical learning theories of the 1950s and early 1960s, the most influential of which was *Stimulus Sampling Theory,* developed by W. K. Estes and colleagues (Atkinson & Estes, 1963; Estes, 1959). Both Stimulus Sampling Theory and adaptive network theory are general classes of learning theories— formal frameworks within which theorists search for a small number of concepts and principles that will illuminate a wide variety of psychological phenomena when applied in varying combinations. To the extent that adaptive networks represent cumulative progress in theory development, we should expect them to incorporate the strengths of Stimulus Sampling Theory but overcome the problems that limited these earlier approaches to modeling associative learning.

This chapter reviews Stimulus Sampling Theory (SST), noting some of its strengths and weaknesses, and compares it to a recent network model of human learning (Gluck & Bower, 1986; 1988a). We will see that the network model's learning rule for updating associative weights represents a significant advance over Stimulus Sampling Theory's more rudimentary learning procedure. In contrast, Stimulus Sampling Theory's stochastic scheme for representing stimuli as distributed patterns of activity can overcome some limitations of network theories that identify stimulus cues with single active input nodes. This leads us to consider a distributed network model that embodies the processing assumptions of our earlier network model but employs stimulus-representation assumptions adopted from Stimulus Sampling Theory. In this distributed network, stimulus cues are represented by the stochastic activation of overlapping populations of stimulus elements (input nodes). Rather than replacing the two previous learning theories, this distributed network combines the best established concepts of the earlier theories and reduces to each of them as special cases in those training situations where the previous models have been most successful.

STIMULUS SAMPLING THEORY

Stimulus Sampling Theory treats learning as a stochastic process in which stimuli are represented as populations of independent variables, called *stimulus elements* (for reviews, see Bower & Hilgard, 1981; Neimark & Estes, 1967). On any individual experimental trial, only a subset of these elements is presumed to be sampled by the subject. Each element in the set is assumed to be completely associated with one of the possible responses available to the subject. Traditionally, two different, but often functionally equivalent modeling schemes have been employed to describe how stimulus elements are sampled. One scheme supposes that each stimulus element has a certain independent probability of being sampled whereas the other scheme assumes that a fixed number of stimuli are taken from the total population on each trial. Once a subset of the population has been sampled, choice behavior is determined by the proportion of sampled elements associated with each response. For example, if 75% of the sampled elements are associated with response R_1 and 25% with response R_2, the model predicts that the subject will respond R_1 with probability .75. Reinforcement occurs through the total conditioning of all sampled elements, each of which becomes associated with the reinforced outcome.

Probability Learning

Many early applications of Stimulus Sampling Theory were concerned with probability learning experiments where subjects were trained to predict which of several randomly chosen outcomes would occur. The most basic probability learning situation involves two possible outcomes, which we refer to as E_1 and E_2. At the beginning of each trial, subjects give one of two possible responses: R_1 if they expect E_1, and R_2 if they expect E_2. The stimulus conditions in this experiment are represented by a single population of stimulus elements. If p_n denotes the proportion of elements connected to R_1 at the beginning of trial n, then p_n is both the probability that a randomly selected element will be connected to R_1 and the probability that the subject will respond R_1 on that trial. If θ is the probability that an individual element is sampled, the theory implies a learning equation described by changes in p_n where

$$p_{n+1} = \begin{cases} (1 - \theta)p_n + \theta & \text{if } R_1 \text{ is reinforced} \\ (1 - \theta)p_n & \text{if } R_2 \text{ is reinforced.} \end{cases} \qquad (1)$$

If the reinforcing event, E_1, occurs with constant probability π, we can rewrite Equation 1 as

$$p_{n+1} = \pi[(1 - \theta)p_n + \theta] + (1 - \pi)(1 - \theta)p_n$$
$$= (1 - \theta)p_n + \theta\pi \tag{2}$$

Following extended training, Equation 2 predicts that p_n, the probability of responding R_1, will come to match π, the objective probability of E_1 occurring. One way to see this is to compute the long-term average proportion of E_1 events (π) and show that this is equivalent to $\left(\frac{1}{T}\sum_{n=1}^{T} p_n\right)$, the long-term average probability of responding R_1 (see Hilgard & Bower, 1975, p. 386). Alternatively, we can rewrite Equation 2 in terms of the expected change in p, Δp, from trial n to trial $n + 1$,

$$\Delta p = \theta(\pi - p_n). \tag{3}$$

Equations 1 and 3 characterize a "linear model" of learning, expressing how p_n, the probability of responding R_1, changes as a linear function of p_{n-1}. From Equation 3 we see that the system will stabilize (i.e., the expected change in $p, \Delta p = 0$) when $p_n = \pi$. This implies that p_n, the proportion of R_1 responses, will come to match π, the average proportion of E_1 events. This appears to be a non-optimal strategy because "probability matching" generally yields a lower expected proportion of correct responses compared to a "probability maximizing" strategy in which a subject always chooses the most likely outcome. This is most easily seen with reference to a two-choice task. A correct response occurs whenever the subject responds R_1 on an E_1 trial (which will occur with probability π^2), or when the subject responds R_2 on an E_2 trial (which will occur with probability $(1 - \pi)^2$). Thus, probability matching yields an expected proportion correct of $\pi^2 + (1 - \pi)^2$. Note that this is always less than or equal to the proportion correct expected from adopting a probability maximizing strategy. For example, if $\pi = .8$, a probability matching strategy results in a long-run average of .68 correct whereas a probability maximizing strategy (choosing R_1) yields a long-run average of .8 correct.

The probability matching prediction of Stimulus Sampling Theory has been tested in a wide variety of training situations and, for the most part, these predictions have been confirmed. For example, Suppes and Atkinson (1960, p. 196) report an experiment in which $\pi = .60$ and the observed proportion of R_1 responses was .596 (averaged for 30 subjects over the last 100 out of 240 trials). The model also makes fairly accurate predictions regarding the shape of learning curves under a variety of reinforcement schedules (Estes, 1964). Other tests of Stimulus Sampling Theory have demonstrated its ability to predict sequential statistics that describe the extent to which a subject's response on trial $n + 1$ is influenced by his responses and/or the reinforcing events on trial n (Atkinson, Bower, & Crothers, 1965; Suppes & Atkinson, 1960). The theory has also been applied, with considerable success, to such diverse phenomena as spontaneous recovery and forgetting (Estes, 1955), reaction time distributions (Bush & Mosteller, 1955), and recognition memory (Bower, 1972).

A   B

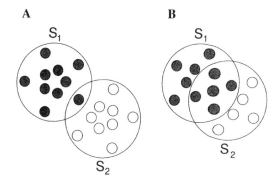

FIG. 9.1. Generalization through shared elements for two stimuli S_1 and S_2 that are: (A) slightly similar (small overlap) or (B) very similar (large overlap).

Stimulus Generalization and Discrimination Learning

In Stimulus Sampling Theory, *stimulus generalization* between distinct stimuli is conceived to arise from common elements shared by both stimuli, an approach drawn from Thorndike's (1898) "connectionism" theory of Stimulus-Response formation. For example, the response associated with stimulus S_1 in Fig. 9.1 will generalize to stimulus S_2 to the extent that they share common stimulus elements. It has long been known that if a stimulus such as a high frequency tone, S_1, is associated with some significant event (R_1), this conditioning will generalize to other similar stimuli such as a low tone (S_2). Within Stimulus Sampling Theory the generalization of the $S_1 \rightarrow R_1$ association to an $S_2 \rightarrow R_1$ association is predicted to be in direct proportion to the amount of overlap between the S_1 and S_2 stimulus pools. This common-elements approach to stimulus generalization has been successfully applied to a range of generalization phenomena (Atkinson & Estes, 1963; LaBerge, 1961).

The application of this approach to *discrimination learning* has, however, been problematic. Consider the case in which stimulus S_1 is paired with response R_1 and stimulus S_2 is paired with response R_2. Stimulus Sampling Theory's learning rule expects the distinctive elements in S_1 and S_2 to become totally conditioned to R_1 and R_2, respectively. The common elements, however, will be conditioned to both responses in weighted proportion to the R_1 and R_2 reinforcement frequencies. Responding to stimuli will therefore be controlled by the correctly conditioned distinct elements as well as the "mixed conditioned" common elements. Thus, the model predicts, incorrectly, that perfect discrimination should never occur. People and animals can, of course, be trained to respond differentially to distinctive stimuli, even when the stimuli are quite similar.

It might seem simple enough to propose a mechanism for discrimination learning whereby the subject learns to "adapt out" the shared common cues and attend only to the unique cues. Several theorists have proposed "attentional" extensions to Stimulus Sampling Theory. One approach is to postulate additional mechanisms that render common elements ineffective during the course of discrimination learning (Bush & Mosteller, 1955; Lovejoy, 1968; Restle, 1957; Sutherland & Mackintosh, 1971). Other approaches include the addition of specialized observing responses (Atkinson, 1958; Levine, 1970) or modified decision processes (Laberge, 1962). One problem with these approaches is that by nullifying the effect of the shared common elements, they predict that these common elements will have no influence on transfer tasks. Several studies have shown that even with complete discrimination training, the common elements can still exert a strong influence on subsequent transfer tasks when subjects are asked to classify novel combinations of cues (Binder & Feldman, 1960; Binder & Taylor, 1969; Flagg & Medin, 1973; Robbins, 1970).

Estes (1959) considered still another approach in which the stimulus situation on a given trial is viewed as a unique pattern rather than as a collection of component cues. This pattern model has a desirable property that the component-cue model lacks: the ability to predict perfect discrimination between two stimulus patterns in the presence of common elements. Mitigating this advantage, however, is the failure of the pattern model to provide an adequate account of stimulus generalization. A natural combination of the component and pattern theories is the mixed model first proposed by Estes and Hopkins (1961) and later developed quantitatively by Atkinson and Estes (1963). According to this model, associations are formed during discrimination training, between the component cues and the responses as well as between the pattern cues and the responses. Once the pattern cues are learned, they are presumed to dominate in discrimination tasks. In generalization tasks, however, the component cues mediate responding. Whereas the mixed model has had some success in resolving the overlap problem, the interactions between the component and pattern processes have not been completely evaluated for the full range of discrimination and generalization tasks. In summary, an entirely satisfactory resolution to the "overlap problem" has not been developed that successfully reconciles stimulus generalization and discrimination learning (Bower & Hilgard, 1981; Medin, 1976).

Binder and Estes (1966): A Stimulus-Sampling-Theory Interpretation

To better appreciate the subtleties involved in trying to resolve the overlap problem in discrimination and generalization, we now consider in detail a study by Binder and Estes (1966). This study illustrates an additional problem with Stimulus Sampling Theory; its inability to account for a phenomena that Binder and Estes termed the "relative novelty" effect. We describe this study (and subsequent extensions and elaborations by other investigators) and then use these data as a test base to compare Stimulus Sampling and adaptive network interpretations of discrimination learning.

Binder and Estes (1966) conducted systematic studies of the effects of category frequency on learning, following a line of research begun by Binder and Feldman (1960). Subjects were trained to classify patterns composed of several simple component cues. Following this training, they classified novel combinations of the component cues. Stimulus patterns ab and ac were reinforced with responses R_1 and R_2, respectively. The critical manipulation was the unbalanced presentation frequencies of the different reinforcements; $ab \rightarrow R_1$ trials occurred three times as often as $ac \rightarrow R_2$ trials.

A Stimulus-Sampling-Theory interpretation of this experiment posits three populations (pools) of elements corresponding to each of the three cues. Presentation of the ab pattern activates elements in both the a pool and the b pool. With the unbalanced presentation frequencies of the two reinforcements, Stimulus Sampling Theory predicts that after extended training, all stimulus elements from the b pool will be associated with R_1, the more common outcome, and all elements from the c pool will be associated with R_2, the less common outcome. The a pool, however, will contain some elements associated with R_1 and others associated with R_2. Because R_1 was presented three times as often as R_2, Stimulus Sampling Theory predicts that 75% of the elements in the a pool will be conditioned to R_1, and 25% will be conditioned to R_2. Therefore, Stimulus Sampling Theory expects that the presentation of symptom a alone should, on average, activate a stimulus sample with the majority of elements predicting the common category. This result accords with data from studies by Binder and Feldman (1960) who used a 2:1 ratio of $ab \rightarrow R_1$ to $ac \rightarrow R_2$ presentations and observed response proportions for the shared common cue (a) of .65 and .29 for R_1 and R_2, respectively (compared to predicted values of .67 and .33). With a 4:1 ratio of presentations frequencies, they observed response proportions for the common cue of .76 and .20 (compared to predicted values of .80 and .20).

As noted earlier, Stimulus Sampling Theory is unable to account for people's ability to discriminate between similar stimuli that activate overlapping populations of hypothetical elements. The same problem exists when the common elements are explicit, as in the Binder and Feldman (1960) and Binder and Estes (1966) studies. As Binder and Estes noted, as long as subjects are randomly sampling elements from the a pool, it is possible that ab patterns will be incorrectly classified with the rare outcome and ac patterns incorrectly classified with the common outcome (p. 3). Not surprisingly, subjects learned to master perfectly the $ab \rightarrow R_1 / ac \rightarrow R_2$ discrimination with relative ease. Because of this shortcoming of the Stimulus Sampling model, Binder and Estes suggested that it might be necessary to augment Stimulus Sampling Theory with a mechanism by which subjects could learn to "respond selectively to cues which are reliable predictors of reinforcing events . . . and to ignore or 'adapt to' common cues . . . which are not uniformly correlated with reinforcement" (p. 4). As described earlier, several schemes for doing just this were proposed in the literature. However, the perfect discrimination attained by subjects in this task would seem to suggest that subjects had learned to "adapt to" or ignore the common a

cue. This is inconsistent with the previously described transfer effect wherein subject's transfer classification of the a cue indicates that they have clearly learned that a predicts R_1, the more common reinforcement. Thus, the use of explicit common elements (cues), as in the Binder and Feldman (1960) and Binder and Estes (1966) studies makes clear the difficulty in reconciling discrimination and generalization behaviors.

The Relative Novelty Effect

Stimulus Sampling Theory also fails to account for another aspect of Binder and Estes' data. During the transfer task, subjects were given the the novel feature combination (bc). Stimulus Sampling Theory expects that early in training bc should be more associated with R_1, the more common outcome, because there will be more b elements associated with R_1 than c elements associated with R_2. Once learning is complete, however, Stimulus Sampling Theory predicts that bc should, on average, activate an equal number of oppositely associated elements, predicting that bc should be equally associated with the two reinforcements. In summary, Stimulus Sampling Theory predicts that bc will be associated either with the more frequent reinforcement (R_1) or with both reinforcements equally. Surprisingly, subjects were more likely to classify the bc pattern with R_2, the *less* frequent outcome. Binder and Estes called this the "relative-novelty" effect because the probability of a stimulus component controlling choice behavior (c in this case) appears to be inversely related to its presentation frequency during training. Subsequent replications and extensions of this result have been presented by Binder and Taylor (1969), Medin and Robbins (1971), and Medin and Edelson (1988). Heretofore, no satisfactory explanation has been offered for it.

GENERALIZATION AND DISCRIMINATION
IN ADAPTIVE NETWORKS

Stimulus Sampling Theory was strongly motivated by the principle that

> . . . we can hope to understand the processes that guide adult human behavior only within a rather broad framework in which they can be meaningfully related both to the more primitive or elementary processes from which they develop during the life of the individual and to those of lower organism. . . . Typically, evolution works through endless variations on a limited repertory of themes . . . as a consequence, clues to understanding complex processes of human cognition sometimes come from studying simpler forms. (Estes, 1982, p. 315)

Subsequent developments in learning theory all but abandoned this unified approach to understanding human and infrahuman learning. Animal research continued to be primarily concerned with elementary associative processes whereas, by the mid-1960s, human learning began to be characterized in terms of information processing and hypothesis testing, concepts borrowed from artificial intelligence and computer science. The recent emergence of "parallel distributed processing" models based on "connectionist" networks, however, presents an alternative to the rule-based symbolic models of the 1970s and early 1980s. Like the earlier statistical learning theories, these network models embody the assumption that many complex human abilities can best be understood as emerging from configurations of elementary associative processes.

In recent papers, we have used adaptive networks to explore the relationship between human learning and the elementary associative learning processes that can be studied in simpler organisms. We began by studying a simple adaptive network model of human learning that extends Rescorla and Wagner's (1972) description of classical conditioning to human classification learning (Gluck & Bower, 1986, 1988a, 1988b; Gluck, Bower, & Hee, 1989). The learning rule is the same as the least mean squares (LMS) learning rule for training one-layer networks, first proposed by Widrow and Hoff (1960). The model has been fit to data from experiments on probabilistic classification learning with multiple cues. Although this simple model can be applied only to a restricted range of experimental circumstances, it has shown a surprising accuracy in predicting human behavior within that range, including data on people's choice proportions during learning, the relative difficulty of learning various classifications, and their responses to generalization tests involving novel combinations of cues. The ingredients of the basic network model are shown in Fig. 9.2A.

Presentation of a stimulus or pattern of cues corresponds to activating one or more of the sensory elements on the left. They, in turn, send their activations to an output unit along associative lines that have amplifier weights, the w_i. The weighted inputs are summed at the output node, and this output $\sum_{j=1}^{n} w_j a_j$, is converted into a response measure. In classical conditioning, the inputs are single to-be-conditioned stimuli such as lights and bells that are paired with the unconditional stimulus, such as food for a hungry dog; the output node reflects the animal's expectation of the unconditional stimulus given the cues presented. In a classification experiment involving human adults as subjects, the stimuli might be patterns of, say, medical symptoms displayed by a patient, and the output reflects the degree to which the model expects such a patient to have some target disease (classification) versus alternative diseases (Fig. 9.2B).

The network operates in a training environment in which reinforcing feedback (the correct classification) is given after each stimulus pattern. The central axiom of the model is its learning rule, which is that the weights, the w_i's, change on each trial according to

$$\Delta w_i = \beta a_i \left(\lambda - \sum_{j=1}^{n} w_j a_j \right) \tag{4}$$

Here, λ is the training signal which might be $+1$ if the category is reinforced (e.g., US present) and 0 otherwise (e.g., no US). The cue-intensity parameter, a_i, is assumed to be 1 if cue i is present on the trial, and 0 if it is absent. The learning

FIG. 9.2. A simple one-layer network that can learn the associations between four stimulus cues and two possible outcomes. (A). The network's classification prediction is a function of the activation on the output nodes. Associative weights between feature nodes and category nodes are updated according to the error-correcting principle of the Rescorla–Wagner (1972) model of classical conditioning, equivalent in this application to Widrow and Hoff's (1960) LMS rule of adaptive network theory. (B). The network applied to a classification experiment involving human adults as subjects, where the stimuli are patterns medical symptoms displayed by a patient and the output reflects the degree to which the model expects such a patient to have some target disease (classification) versus alternative diseases.

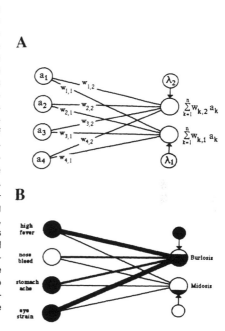

rate, β, is a parameter (on the order of .01 in most simulations) that determines how much the weights change when the output differs from the training signal, λ. Equation 4 is variously called the delta rule, the least-mean-square (LMS) rule, or the Rescorla–Wagner conditioning rule (cf. Sutton & Barto, 1981).

Comparing Network and SST Learning Rules

It is instructive to compare Equation 4 of the LMS rule to Equation 3, the linear operator rule from Stimulus Sampling Theory. If we identify p in Equation 3, the probability of responding R_1 with w_1, θ with β, we can re-express the linear operator rule in the terminology of adaptive networks as

$$\Delta w_1 = \beta(\lambda - w_i). \tag{5}$$

Comparing the linear operator rule (Equation 5) of Stimulus Sampling theory with Equation 4 of the Rescorla–Wagner/LMS rule, we note one key difference. Weight changes in the Rescorla–Wagner/LMS rule are governed by the difference (or discrepancy) between the reinforcement (λ) and the network's expectation of the reinforcement, $\sum_{j=1}^{n} w_j a_j$ (the output), which is sensitive to all the cues present on a trial. In contrast, Stimulus Sampling Theory operates on each cue independently; weight changes depend only on the difference between the reinforcement and the current association between cue i and the reinforcing outcome. Note that in training situations where individual component cues are present as complete patterns (as in probability learning studies), Equation 4 of the LMS rule reduces to Equation 3 of Stimulus Sampling Theory. Thus, it is only in training procedures involving patterns of multiple cues that have the opportunity to "compete" among themselves to reduce the error, $(\lambda - \sum_{j=1}^{n} w_j a_j)$, will we expect to see divergent predictions from SST and the LMS network.

Stimulus Sampling Theory follows the tradition of Hull (1943) and Spence (1936) in assuming that the temporal contiguity, or joint occurrence, of a cue and a reinforcing outcome is sufficient for associative learning. This view, however, came under serious attack in the late 1960s, just as interest in Stimulus Sampling Theory began to wane. The work of Kamin (1969), Rescorla (1968), and Wagner (1969) demonstrated that the ability of a previously neutral conditioned stimulus (CS) to become conditioned to an unconditional stimulus (US) depends on the CS imparting reliable, nonredundant, and predictive information about the expected reinforcement. For example, in Kamin's (1969) "blocking" experiment, a light, the CS, was first conditioned to predict a shock, the US. In a subsequent training phase, a compound stimulus consisting of a light and a tone was paired with the shock. Surprisingly, learning of the *tone* → *shock* association hardly occurred at all compared to control subjects who had received no pretraining to the light. One interpretation of blocking and related effects is that animals are learning to modulate the processing of sensory cues in order to adapt out (ignore) the irrelevant cues such as the tone in the given example (Mackintosh, 1975; Pearce & Hall, 1980). These explanations are reminiscent of extensions to Stimulus Sampling that sought to reconcile stimulus generalization with discrimination learning (e.g., LaBerge, 1962; Restle, 1957). Kamin (1969) suggested an alternate interpretation of these attention-like effects. He proposed that the blocking effect results not from modulation of CS-processing but rather from modulation of US-processing. If the effectiveness of a US for producing associative learning depends on the relationship between the CS and the *expected outcome*, little additional learning would occur once the animal had already learned to anticipate (predict) the US (Kamin, 1969).

Rescorla and Wagner provided a precise formulation of Kamin's proposal (Rescorla & Wagner, 1972; Wagner & Rescorla, 1972) and it is this rule that we employ to train the weights in our adaptive network model of human learning. Rescorla and Wagner's conditioning model assumes that the association between a stimulus and its outcome changes on a trial proportional to the degree to which the outcome is unexpected (or unpredicted) given *all* stimulus elements present on that trial (Equation 4). The Rescorla–Wagner model accounts for the blocking effect as follows: When in Phase 1, CS_1 has been initially conditioned to the US, w_1 approaches 1 (assuming $\lambda = 1$ for US trials). If the initial associative strength of the novel stimulus, w_2, is zero, then the compound stimulus strength, $w_1 + w_2$, will already equal 1 at the beginning of Phase 2. By Equation 4, the incremental change in the associative weight of both stimuli is predicted to be zero when the compound is paired with the US during Phase 2. In contrast to cue-adaptation theories that assume that "attentional" phenomena are mediated by variations in CS processing. Rescorla and Wagner showed how many of these same phenomena could be more readily understood as resulting from variations in US processing.

LMS and the Overlap Problem

Turning back to the "overlap" problem of Stimulus Sampling Theory, we see that the Rescorla–Wagner/LMS rule provides a mechanism for effectively adapting out common irrelevant cues. Consider two stimulus patterns, $P\,1$ and $P\,2$, that are represented by distinct populations of stimulus elements, S_1 and S_2, as well as a common population, S_c. If S_1 is associated with a reinforcing event, R_1, associative strength will accrue to both S_1 and S_c. This association will generalize to $P\,2$ via the overlapping elements in S_c that are shared with $P\,1$.

In a discrimination training procedure, however, $P\,1$ might be associated with R_1 and $P\,2$ with R_2. One possible network representation of this problem is to have a single output node that receives a training signal of $+1$ when R_1 is reinforced and a training signal of -1 when R_2 is reinforced. Under these conditions the competitive learning principle of the Rescorla–Wagner/LMS rule will seek a solution whereby $w_1 + w_c = +1$ while $w_2 + w_c = -1$. One possible solution is to have all of the associative strength accrue to $w_1 = +1$ and $w_2 = -1$, with $w_c = 0$ "adapting out" so that the system achieves errorless discrimination (see also Rudy & Wagner, 1975, p. 290). As we see later, however, it is possible under some training procedures for the LMS network to find other solutions that do not require that $w_c = 0$. A major challenge for the LMS network—and all models of learning—is to try and reconcile the role of common elements in both stimulus generalization and discrimination learning.

LMS and Probability Matching

Like Stimulus Sampling Theory, the LMS network will generally predict probability matching in choice behavior when the output activations (or a mono tonic transformation of them) are converted to choice probabilities using a likelihood ratio rule (Gluck & Bower, 1988a). The relationship between the Least Mean Squares solution and probability matching can most easily be seen with reference to a single output node that is reinforced ($\lambda = 1$) with probability π. If A is the output activation of the node, then the squared error will be $(1 - A)^2$ with probability π and A^2 with probability $(1 - \pi)$. Thus, the expected mean squared error (MSE) is

$$E[MSE] = \pi(1 - A)^2 + (1 - \pi)A^2. \tag{6}$$

To find the value of A that minimizes the expected mean squared error, we differentiate Equation 6 with respect to A:

$$\frac{d(E[MSE])}{dA} = -2\pi(1 - \pi) + 2(1 - \pi)A. \tag{7}$$

By setting $d(E[MSE]) = 0$ we solve for A to find that the minimum squared error occurs when $A = \pi$ (see also Gluck & Bower, 1990, p. 108). Thus, we expect that the LMS algorithm will converge to a set of weights that result in the closest possible approximation to having the output activations reflect the observed probabilities of reinforcement for each pattern in the training set. Because the approximation to probability matching on the output activations is not necessarily bounded between 0 and 1 (as are the p in Stimulus Sampling Theory) it may be necessary to transform the network's output activations before mapping them onto expected choice probabilities. Examples of using the LMS network to fit observed data on probability matching can be found in Gluck and Bower (1988a), Estes, Campbell, Hatsopoulos, and Hurwitz (1989), and Shanks (1989).

Binder and Estes (1966): An Adaptive-Network Interpretation

We return now to the Binder–Estes study to see what the LMS network predicts here. Medin and Edelson (1988), in their replication and extension of the Binder–Estes study, noted that the "relative-novelty" effect is qualitatively consistent with competitive learning rules, such as the Rescorla–Wagner rule. Their logic goes as follows: Assume that cues a and b compete to predict the common category while cues a and c compete to predict the rare category. Because a occurs more often with the common rather than the rare category, it will presumably acquire more associative weight to the common category. Thus, a will compete with b to predict the common category, thereby diminishing b's association to the common category. For pattern ac to predict the rare category, symptom c will have to overcome the association of a to the common category. This leads us to expect that when b and c are paired together, c's association to the rare category should be stronger than b's association to the common category. Thus, a competitive-learning principle might expect that the novel test pattern bc should be judged more strongly associated with the rare category, as observed by Binder and Estes (1966) and Medin and Edelson (1988).

Given this reasoning, we might expect that the LMS network model in Fig. 9.2, which incorporates Rescorla and Wagner's competitive learning rule, should account for the relative-novelty effect. However, as Medin and Edelson (1988, p. 75) note the Rescorla–Wagner model predicts that with extended training, b and c will accrue all the associative strength, leaving a with none. Figure 9.3 shows an adaptive network model of the Binder–Estes/Medin–Edelson experiment. The network has three input nodes: one for each of the three symptoms. All

FIG. 9.3. An adaptive network model of the Binder–Estes task using a local representation scheme with one input node for each of the three stimulus components and a single output node that receives reinforcement of +1 for the more

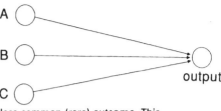

common outcome and −1 for the less common (rare) outcome. This one-output node model is equivalent in this application to a two-output-node model in which each output nodes receives reinforcement of 1 (category present) or 0 (category not present).

weights are initialized at 0. The presence or absence of cue-i is represented by an input node activation, a_i, of 1 or 0, respectively. The output node is reinforced with $\lambda = +1$ for the common category and $\lambda = -1$ for the rare category. This one-output-node model, with $+1/-1$ reinforcements, yields identical predictions to a two-output-node model with $1/0$ reinforcements where each output node corresponds to one of the possible outcomes (see Gluck & Bower, 1988a, footnote 2, p. 234, for more details on this correspondence).

Figure 9.4 graphs the changes in weights for the three input nodes (cues) during training, the output activations for the training patterns during learning, and the output activations (responses) for the transfer patterns at each stage in learning. These simulations are from a network run for 200 trials with a learning rate, β, of .03; so long as β is sufficiently small, however, the important ordinal predictions of the model are independent of the particular parameter value chosen. The simulation in Fig. 9.4 confirms Medin and Edelson's observation that extended training with the Rescorla–Wagner/LMS rule results in cues b and c acquiring all the predictive strength: asymptotically, $w_B = +1$, $wc = -1$, whereas cue a adapts out, with $w_A = 0$. Thus, b is completely associated with R_1, the common category, c is completely associated with R_2, the rare category, and a has no associative strength at all. As shown in Fig. 9.4A, a does acquire a pre-asymptotic association to the common category (i.e., a positive weight). Thus, the network's early response to pattern a is consistent with both Binder and Feldman's (1960) and Medin and Edelson's (1988) results on the common-cue test (a). Medin and Edelson (p. 75) and subsequently we (Gluck & Bower, 1988a) incorrectly suggested that the relative-novelty effect will also emerge from the Rescorla–Wagner model as a pre-asymptotic effect. As the simulations in Fig. 9.4 demonstrate, this is clearly incorrect. Figure 9.4C shows that the network's response to the transfer pattern, bc, favors the common category at all stages of learning prior to asymptotic learning. Because of the imbalance in presentation frequencies, the response to bc remains positive despite a's transient association to the common category, because w_b increases in strength towards $+1$ much faster than the w_c approaches -1. Early in training, during the tran-

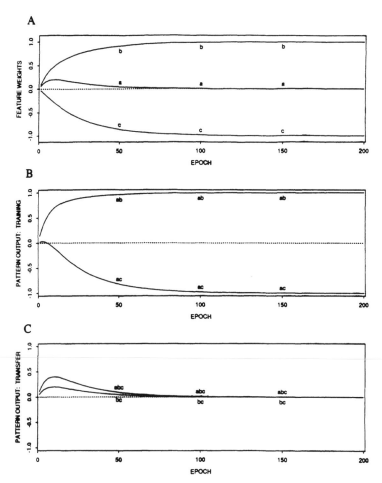

FIG. 9.4.   Gluck and Bower's adaptive network model applied to Bind-er–Estes/Medin–Edelson experiments. (A). Changes in weights for the three input nodes (cues) during training across training. Positive weights and activations favor the common category whereas negative weights and activations favor the rare category. (B). Output activations for the training patterns during training. (C). Predicted responses to transfer tests at each point in training. In all these figures, positive weights and activations favor the common category whereas negative weights and activations favor the rare category.

sient association of a to the common category, b's association to the common category is always greater than c's association to the rare category. Thus, at no time during the course of training will the Rescorla–Wagner model—or, equivalently, the adaptive network model of Gluck and Bower (1988a)—predict the "relative novelty" effect on the bc test.

Understanding the Network's Solution

Why does the network's behavior in this training situation differ from our intuitive expectation of what a competitive learning rule should do? When the network is trained as just described, the LMS rule converges on the "solution vector," $W_{[a,b,c]} = [0, +1, -1]$. Note that this is only one of many possible solution vectors that would be equally effective in solving the ab/ac discrimination. For example, if $W_{[a,b,c]} = [.2, .8, -1.2]$, this would also result in errorless performance. In a deterministic task that can be perfectly solved by the network (i.e., MSE = 0), the set of solution vectors is unaffected by variations in the presentation frequencies of the individual training patterns. This type of problem, for which multiple solutions exist, can be contrasted with other discrimination problems that have unique solutions. For example, the nondeterministic classification task in Experiment 1 of Gluck & Bower (1988a, Appendix A) has a unique solution that can be derived analytically. When a unique network solution exists, the LMS algorithm will converge on that solution independent of the initial weights, assuming a sufficiently small learning rate (Widrow & Hoff, 1960). In situations where multiple solutions exist, such as the Binder–Estes/Medin–Edelson task, the final weights obtained with the Rescorla–Wagner/LMS algorithm will be sensitive to their initial values (Gluck & Bower, 1988b, Appendix B; Parker, 1986). The sensitivity of the LMS rule to initial conditions is familiar to animal learning theorists as the property that allows the Rescorla–Wagner model to account for the effect of pretraining in Kamin's (1969) blocking study.

If many different solutions are equally "good" in minimizing the expected squared error, why does the network converge to $[0, +1, -1]$ rather than another solution, for example, $[.2, .8, -1.2]$? To see why, it is helpful to consider the set of all possible solution vectors as being a subset of the three-dimensional "weight space" that characterizes all possible states of the three-weight network. If the network begins with all weights set to zero, then the solution with the smallest sum squared weights represents the "closest" solution to the initial conditions, where closeness is measured by Cartesian distance. Parker (1986) has shown that if the weights in the network are initialized at zero (or randomly distributed with zero mean), the asymptotic weights will tend toward the solution closest to the initial conditions. For the network model of the Binder–Estes/Medin–Edelson task we expect, on average, a solution where cues "b" and "c" have all the weight because the solution to the simultaneous linear equations $w_A + w_B = +1$ and $w_A + wc = -1$, with the smallest sum-squared weights is $w_A = 0$, $w_B = +1$, $wc = -1$. (See Gluck & Bower, 1988b, Appendix B, for more details on deriving the expected asymptotic convergence when multiple solutions exist).

STIMULUS SAMPLING AND THE RESCORLA-WAGNER MODEL

It is clear from the analyses presented that neither Stimulus Sampling Theory nor our LMS network model provides an adequate account of the effects of category frequency on discrimination learning and transfer generalization when the training patterns share common cues. Stimulus Sampling Theory accounts best for transfer effects involving single component cues. This "component matching" principle was summarized by Binder and Tayler (1969) as: "If two or more different responses have been reinforced in the presence of a given cue during training, then with any later tests the probability that any one of these responses will be evoked by the given cue is equal to its relative frequency of reinforcement" (p. 91). In contrast, adaptive network theory, and the Rescorla–Wagner/LMS rule in particular, provides a better account of discrimination learning when the reinforcement contingencies for cues are dependent on the context in which they appear. The LMS rule converges on a set of weights that (as closely as possible) produce "pattern matching" probabilities as activations on the output nodes when the individual cue weights are combined additively. To paraphrase Binder and Taylor: The LMS network seeks a solution whereby if two or more different responses have been reinforced in the presence of a given *pattern* during training, then with any later tests the output activation evoked by the given *pattern* for one of these responses will be equal to its relative frequency of reinforcement.

A possible rapprochement between the explanatory abilities of Stimulus Sampling Theory and the Rescorla–Wagner model has been suggested by Rescorla (1976) and Blough (1975). These authors have shown that integrating the learning rule from Rescorla–Wagner's (1972) conditioning model with the stimulus-representation assumptions from Stimulus Sampling Theory can account for several animal learning behaviors. We first review the applications of this hybrid model to animal learning behavior and then consider its implications for human classification learning.

Rescorla (1976)

Rescorla (1976) highlighted an important implication of the Rescorla–Wagner model when it is applied to a distributed representation in which similar stimuli share common features. Traditional learning theory says that the optimal way to train an associative connection is by direct reinforcement of the target stimulus. Rescorla, however, reported a paradoxical case where training to a *generalized* stimulus enhances performance to a target stimulus more than reinforcement to the target stimulus itself. This paradoxical outcome and the circumstances that produce it are predicted by Equation 4 of the Rescorla–Wagner/LMS rule.

If we conceptualize the similarity of two stimuli in terms of common or shared stimulus elements, then we may represent even a simple stimulus like a high-pitched pure tone as a compound of stimulus elements, denoted AX; another similar stimulus like a low-pitched pure tone would be represented as another compound, BX. Here, X denotes the set of common elements, whereas A and B denote those sets of sensory elements unique to the two stimulus sets. In conditioning the high tone (AX) to a shock US, Equation 4 applies to the separate A and X components of that stimulus. With repeated reinforcement of AX, the weights w_A and w_X will increase together until their sum equals the reinforcing value of $\lambda = 1$ where each might be, say, about ½. Were we to continue to reinforce AX beyond this point, the LMS rule of Equation 4 expects no change in the strengths of the A and X associations. Now consider what would happen if we gave trials wherein a generalized stimulus (the low tone, denoted BX) was paired with shock. Because B begins at low strength, the combination BX begins with an association strength far below 1. During a block of reinforced trials on BX, Equation 4 implies that w_B and w_X will increase. This increase in w_X should be most apparent when we test the subject again on the original training stimulus, AX. On such a test, the compound strength $w_{AX} = w_A + w_X$ will be higher than before, higher even than if the subject had just continued training on AX alone. In an experiment of this kind, Rescorla (1976) found just this result. Training to a generalized stimulus (following initial learning) produced greater conditioned responding to a target stimulus than did extended training on the target stimulus itself. This is a most counterintuitive result, and one that provides impressive support for the Rescorla–Wagner learning rule when combined with a common-elements representation of stimulus similarity.

Blough (1975)

Nearly concurrently with Rescorla (1976), Blough (1975) described a stimulus sampling model that incorporated a generalization of the Rescorla–Wagner rule very similar to Widrow and Hoff's (1960) LMS rule. Blough assumed that a stimulus continuum (such the pitch of a tone or the wavelength of light) could be represented as an ordered sequence of overlapping sets of hypothetical stimulus elements. Presentation of a physically-defined stimulus corresponds to sampling a subset of these elements according to a unimodal bell-shaped probability distributed with mode at the internal unit corresponding to the physical stimulus. Thus, different physical stimuli are assumed to project probabilistically to internal sensory units that overlap to varying degrees, as illustrated in Fig. 9.5.

FIG. 9.5. Probability distributions for sampling sensory units
given the presentation of physical stimuli at three levels along
a stimulus continuum.

Ordered
sensory
units

Consider now what happens if stimuli below some point on the continuum are reinforced at one rate (e.g., on one twelfth of the trials) while stimuli above that point are reinforced at another rate (e.g., on one third of the trials). What does the Rescorla–Wagner/LMS rule predict will happen with this differential training? The predicted asymptotic association strengths for each stimulus element are shown as a dashed line in Fig. 9.6A. As expected, the model predicts that subjects should adjust their conditional associations to reflect the two reinforcement levels. The distributed representation of the physical stimuli results in a smooth transition from one level of responding to the other because elements nearby the transition point are activated by physical stimuli both above and below the transition point. This smooth transition is also predicted by the linear operator rule from SST when applied to the same stimulus representation.

More interesting than the gradual transitions from one level of responding to the other are the exaggerated "shoulder" and "trough" on either side of the transition point in Fig. 9.6A. These predicted contrast effects are analogous to edge-enhancement (so-called "Mach bands") in sensory psychophysics. The unique prediction of the Rescorla–Wagner model is that the elements just a little bit further away from the transition point should become "superconditioned" because they frequently co-occur—and hence compete for associative strength— with near-edge elements whose associative strengths reflect conditioning at mixed levels of reinforcement.

The plausibility of such contrast effects in discrimination learning were demonstrated by Blough (1975) who trained hungry pigeons to peck a colored key for food. Keypecks at wavelengths below 597nm were reinforced on one twelfth of the trials, whereas keypecks at wavelengths above 597nm were reinforced on one third of the trials. As predicted by Blough's Stimulus-Sampling extension of the Rescorla–Wagner model, the animals' pecking rates showed marked shoulder and trough effects whereby the cues that were just above and below the transition point for reinforcement (597 nanometers) appear to be "superconditioned" beyond the steady-state levels associated with wavelengths further away from 597 nanometers (Fig. 9.6).

A DISTRIBUTED 'STIMULUS SAMPLING'
NETWORK MODEL

An adaptive network model in which stimuli are represented by stochastically activating overlapping populations of input nodes (stimulus elements) is one type

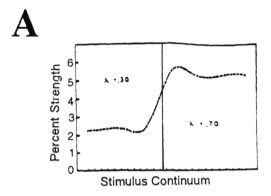

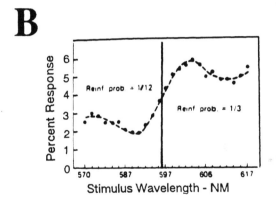

FIG. 9.6. (A). Predicted asymptotic strengths from the Rescorla–Wagner rule for discrimination training on a stimulus continuum. (B). Response rates of three pigeons to key lights varying in wavelength. The vertical line separates the high and low reinforcement stimuli. Note the trough and peak to the left and right of the edge. (From Blough, 1975).

of "parallel distributed network" (McClelland & Rumelhart, 1986; Rumelhart & McClelland, 1986). Networks like this that embody a *distributed representation* use entire patterns of activation across many units to represent different concepts, with different patterns of activation corresponding to different concepts or features (Hinton, McClelland, & Rumelhart, 1986). This is in contrast to a *local representation*, in which each unit (node) in the network is taken to represent a single concept or feature (Feldman, 1985). This "local" representation is what we have previously adopted in our models of human learning, as illustrated in Fig. 9.2B.

A key feature of distributed representations is that each unit is involved in representing many different concepts. Whereas local representation schemes are conceptually easier to understand, many network theorists have been led to adopt distributed representations because of some of their interesting emergent properties. The most compelling advantage of a distributed representation is its ability to generalize automatically from previous training to similar novel situations. As in the earlier Stimulus Sampling Theory, generalization emerges in distributed representations because similar conceptual entities are encoded by activating overlapping sets of units. Several researchers have used this property of distributed networks to account for various aspects of cognitive functioning (Anderson, Silverstein, Ritz, & Jones, 1977; Hinton & Anderson, 1981). Another appeal of distributed representations is the compelling intuition that they are more biologically plausible than local representation schemes (Lashley, 1929; Sejnowski, 1988; Thompson, 1965), but we focus here solely on the behavioral implications of distributed representations.

Although many of the generalization properties of distributed representations bear a marked resemblance to human behavior, there has been little attempt to apply these principles to fitting precise details of human learning and generalization. Part of the problem has been that there is little consensus among theorists as to how external stimuli should be identified with distributed patterns of activity. Building on the previous successes of Stimulus-Sampling-Theory's stimulus representation provides a possible formalism for developing a "distributed" network theory of psychology representation.

Binder and Estes (1966): A Distributed-Network Interpretation

Incorporation of Stimulus Sampling into the network model requires consideration of two new factors. First, it adds an element of randomness or stochasticity to our representation of the stimulus conditions operating on a trial. In the previous "local" network model (Gluck & Bower, 1988a, 1988b), the functional representation of stimuli was identical to the nominal stimuli as described by the experimental paradigm. In stimulus-sampling, the functional representation is presumed to include only a random subset of the nominal stimulus conditions. The second factor introduced by a sampling representation is the explicit incorporation of stimulus similarity through the activation (sampling) of common stimulus elements.

To better understand the unique implications of these two factors, we begin by considering a "non-overlapping" model, in which the population pools for the three cues, a, b, and c are distinct and have no common elements. The behavior of this model will address the implications of adding stochasticity to our model, independent of the effects of common-elements. After analyzing the behavior of this non-overlapping model, we consider an "overlapping"model, which incorporates stimulus similarity among the three cues through the activation of common elements.

Stochasticity in Stimulus Representation

We begin by assuming three distinct pools of input nodes, each having n elements (Fig. 9.7). Presentation of a stimulus cue is presumed to stochastically activate the elements in the corresponding pools with probability θ. Thus, on average, we can expect $n\theta$ elements to be activated in each population of elements. Figure 9.8A shows the result of this non-overlapping network model applied to the Binder–Estes/Medin–Edelson experiments. In Fig. 9.8C we see that the response to the transfer pattern bc has an initial upward swing due to the unequal presentation frequencies of the two categories. At asymptote, however, the associative strength of c for the rare category outweighs the strength of b for the common category and the model correctly predicts that the compound conflicting test, bc, will favor the rare category. Thus, the addition of a *stochastic sampling processes* to the network provides a formal instantiation of Medin and Edelson's qualitative proposal for how a competitive learning rule can account for the "relative novelty" effect.

Why does the relative novelty effect emerge in these simulations and not in a "local representation" network? Note that the formal properties of the network have not changed. What has changed is only our assumption about how the external world is represented on the input nodes. If the sampling rate θ is set to 1, the distributed model reduces to the local model. The important consequence of

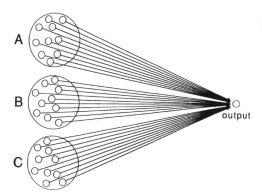

FIG. 9.7. An adaptive network model of the Binder–Estes task using a representation scheme with one pool of input nodes for each of the three stimulus components.

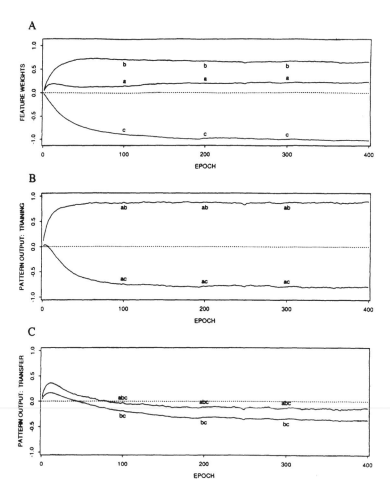

FIG. 9.8. A distributed network model applied to the Binder–Estes/Medin–Edelson experiments with β = .03, n = 20 and θ = .2. (A) shows the expected summed weights (for a given trial) for weights from each of the three input nodes (cues) during training. Positive weights and activations favor the common category whereas negative weights and activations favor the rare category. (B) graphs the expected output activations for the training patterns during training. (C) shows the expected responses to transfer tests at each point in training. Parameter values used were: β = .03, n = 20, and θ = .2. As in Fig. 9.4, positive weights and activations favor the common category whereas negative weights and activations favor the rare category. Each graph is the result of averaging many simulations of these conditions.

the change in representation is that it alters the nature of the discrimination from a deterministic problem to a probabilistic problem. With a stochastic representation of the stimulus environment, there exists a unique set of weights that provide a least mean squares solution. In situations where there is a unique solution, the distributed and local network models will make effectively equivalent predictions. For example, we have analyzed the stimulus environments presented in Experiments 1 and 2 from Gluck and Bower (1988a), and the important ordinal predictions of the original model are maintained by the sampling network. A more detailed exposition of the translation from local to distributed representations, and the conditions under which they make equivalent predictions, can be found in Stone (1986).

Because the processing assumptions of this SST-Network model are identical to those of the previously described network, we can use the same analytic tools for deriving asymptotic solutions (see Appendix A, Gluck & Bower, 1988a). To do so requires making assumptions about n and θ (β has no effect on the expected asymptotic weights). With the values of $n = 20$ and $\theta = .2$ used in the simulation in Fig. 9.8, the expected asymptotic weights for the individual nodes in the three pools are $w_A \approx .057$, $w_B \approx .160$, $wc \approx -.256$. An average of $n\theta = 4$ nodes will be active on each trial in each pool, so the expected activations resulting from the presence of each of the component cues alone is .227, .644, and -1.02, for a, b, and c, respectively.

An intuitive explanation of the relative novelty effect is that the sampling representation adds an additional constraint to the search for an appropriate solution. The constraint can be loosely characterized as *robustness* and is a direct consequence of introducing "noise" into the training procedure. The network now searches for a solution that not only solves the $ab \rightarrow R_1/ac \rightarrow R_2$ discrimination, but is also maximally tolerant of noisy information about cues. For example, if we knew that the first feature was a but were unsure about the second feature, we would want the network to prefer the common category. The Rescorla–Wagner/LMS rule, by virtue of its competitive nature, searches for a parsimonious solution in which redundant information is ignored. But a parsimonious solution may be brittle in that it requires complete and perfect information about all features in the stimulus pattern. When multiple solutions are equally valid, the addition of stochastic noise biases the system toward solutions that are noise tolerant. Similar results have been found for learning in multilayer adaptive networks (Elman & Zipser, 1988; Hanson, 1990). We noted earlier that the LMS rule converges on weights that approximate probability matching on stimulus *patterns* in contrast to the linear operator rule from SST that converges on weights that result in probability matching on *component cues*. The addition of stochasticity to the network provides an additional constraint: When multiple solutions exist that are equally effective in producing "pattern matching," the network will prefer the solution that best approximates "component matching."

The addition of the stochastic sampling representation changes a deterministic discrimination into a probabilistic discrimination in which the expected squared error can never be totally reduced to zero. This does not imply, however, that choice performance, as measured by expected percent correct, cannot reach 100%. In contrast to Stimulus Sampling Theory, which directly maps associative weights onto response probabilities, the network model is not committed to a specific response mapping rule. Rather, it assumes that an unspecified monotonic rule converts activations into response probabilities. For example, in Gluck and Bower (1988a,b), we adopted the sigmoidal transform, which has one free parameter describing the gain or slope of the "S-shaped" sigmoid. As long as all input patterns are more associated with their correct response than with any incorrect response, the sigmoidal response rule can bring choice performance arbitrarily close to 100%, depending on the gain constant. Even with a more moderate gain constant, the sigmoidal transform has the effect of compressing activations near the boundary values so that output activations from all large weights are near unity.

Limitations of the Non-overlapping Representation

One problem with the predictions in Fig. 9.7 is that the model fails to account for Medin and Edelson's finding that subjects, under some conditions, judge the pattern *abc* to be more strongly associated with the common category. We now consider the effects of incorporating stimulus overlap, as well as stochasticity, in our representation of the stimulus cues and show how this effects the predicted response to the *abc* pattern.

With three cues, S_A, S_B, and S_C, we could model the overlapping elements shared by S_A and S_B, S_A and S_C, S_B and S_C, and those shared by all three cues as shown in Fig. 9.9. A simpler alternative, which we adopt here, is to assume that presentation of a cue results not only in the activation of nodes in its "own" pool (with probability θ), but also in the activation of nodes in other pools (with probability θ').

As θ' approaches θ, the weights (and output activations) become more biased towards the presentation frequencies of the categories. Note that in the extreme case in which $\theta = \theta'$, the network has no discriminative input, only random noise on the input nodes, and the input activations would be uncorrelated with the stimulus environment. In this case, the distributed network model reduces, in all aspects, to the Stimulus Sampling Theory account of probability matching described earlier. Clearly $\theta' \ll \theta$ if the cues are reasonably discriminable. Choosing the same values for θ and n that we used in the simulation in Fig. 9.8, and setting $\theta' = .07$, the results of the new simulation are shown in Fig. 9.10. By comparing Fig. 9.10C with Fig. 9.8C, we see that the incorporation of a common-elements representation of stimulus similarity alters the model's behavior so that the correct predictions are obtained for both the relative-novelty test ($bc \rightarrow R_2$) and the combined test ($abc \rightarrow R_1$). The degree of overlap, $\dfrac{\theta'}{\theta}$, will determine the relative strength of this combined test result.

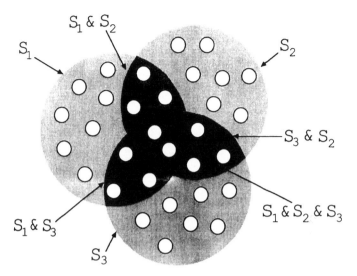

FIG. 9.9.   One method for having three overlapping pools of stimulus elements corresponding to three stimulus cues, S_1, S_2, and S_3.

The model also predicts that both the *bc* and *abc* patterns should show a preference for the more common outcome (R_1) very early in training, with the *bc* pattern reversing its preference with further training. No data is currently available, however, to test this "crossover" prediction.

In comparing our account of the classification of the relative-novelty (*bc*) and the combined patterns (*abc*), it is important to note that the explanation of the former is parameter free $(0 < \theta < 1)$. Thus, we expect the "relative novelty" effect to be strong and reliable. Indeed, it has been replicated by various investigators over the last 20 years (Binder & Estes, 1966; Binder & Taylor, 1969; Medin & Edelson, 1988; Medin & Robbins, 1971). In comparison, our account of the combined test (*abc*) depends critically on the "stimulus confusability" of the component features as measured by the magnitude of θ' relative to θ. Here we can only make the weaker claim that the model is sufficient to account for the *abc* $\rightarrow$ common preference. However, this dependence on the relative magnitude of θ' suggests that experimental manipulations designed to influence stimulus confusability might change the results of the combined test. One such manipulation was used by Medin and Edelson in their fourth experiment. They gave subjects one of two types of instructions. In the <u>focus</u> condition, subjects were told to focus on the symptoms that proved most reliable. In the <u>complete</u> condition subjects were told that they should learn about all of the symptoms. If we assume that the focus condition decreased the opportunity for stimulus confusion, effectively lowering θ', then we expect subjects in the focus condition to exhibit a stronger "relative novelty" effect on the conflicting test (*bc*), but less bias for the

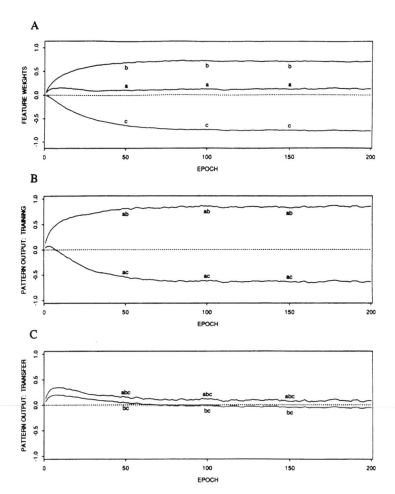

FIG. 9.10. A distributed network model applied to the Binder–
Estes/Medin–Edelson experiments with $\beta = .03$, $n = 20$, and sampling
probabilities of $\theta = .2$ and $\theta' = .07$. (A) shows the expected summed
weights (for a given trial) for weights from each of the three input
nodes (cues) during training. Positive weights and activations favor
the common category whereas negative weights and activations favor
the rare category. (B) graphs the expected output activations for the
training patterns during training. (C) shows the expected responses to
transfer tests at each point in training.

common category on the combined (*abc*) tests. This is precisely what Medin and Edelson found. On the conflicting tests, the focus group showed a considerably larger bias for the rare category than did the subjects in the complete group. On the combined test, subjects in the complete condition exhibited the same effect found in the other experiments—they showed a bias for the common category. Subjects in the focus group, however, showed the opposite effect, exhibiting a slight preference for the rare category versus the common category. Thus, when given the "focus group" instructions, subjects behaved much like the network in Fig. 9.8 with an effective θ' of 0.

Summary

Binder and Estes (1966) were able to account for some of their data within the framework of Stimulus Sampling Theory. They were unable, however, to account for two phenomena. First, subjects discriminate perfectly patterns that share common features. Second, subjects exhibit a relative novelty effect in which two conflicting cues are more strongly associated with a less frequent outcome. Following Rescorla (1976) and Blough (1975), we have developed a distributed network model that combines the learning rule from Rescorla and Wagner's (1972) conditioning model with the stimulus representation assumptions from Stimulus Sampling Theory. This distributed network predicts Binder and Estes' relative-novelty effect (*bc* → less frequent outcome), component matching on transfer (*a* → more frequent outcome), and provides a possible account for Medin and Edelson's (1988) combined-pattern results (*abc* → either, depending on training instructions).

These analyses suggest that it may not be appropriate to apply the Rescorla–Wagner/LMS rule directly to a veridical representation of stimuli, especially for deterministic discriminations. In discrimination tasks in which an infinite number of least-squares optimal solutions exist, the "local" network will often find a solution that is simple and nonredundant but extremely sensitive to "noise" or perturbation. These solutions do not yield transfer generalizations in accord with empirical data. Training the network with a stochastic representation of the input stimuli results in a "robust" solution that generalizes more effectively. Stimulus Sampling Theory provides a stochastic formalism for representing input stimuli.

Configural Cues and Stimulus Sampling

In these preliminary analyses of a distributed network model, we have represented stimulus patterns as collections of component cues (e.g., *a*, *b*, and *c* in the Binder–Estes/Medin–Edelson experiments). In other work (Gluck, in press; Gluck, Bower, & Hee, 1989) we have extended this component representation to include pair-wise conjunctions of features as unique cues. This "configural-cue" model accounts for several aspects of complex human category learning and animal learning. We have not included these "higher order" cues in our analyses of the Binder and Estes and Medin and Edelson experiments, because the relevant configural-cues, *ab* and *ac*, are redundant in this application with the component-cues, *b* and *c*. The consideration of configural-cues does not change the basic predictions of the network model for the Binder–Estes/Medin–Edelson discrimination task. In general, however, the configural-cue approach is perfectly compatible with a distributed stimulus-sampling representation. Configural-cues, like component-cues, can be represented as populations of elements. Furthermore, the geometric interpretation of stimulus similarity as shared common elements (Figs. 9.1 and 9.9) can be extended to represent configural-cues. When overlapping elements are activated only by the presence of both cues (rather than by either alone), they represent unique configurations of the component cues, that is, configural-cues.

CONCLUSION

In describing the strategy for theory development that guided the growth of Stimulus Sampling Theory, W. K. Estes (1982) wrote:

> The approach I favored [was] starting with associative concepts already established for simpler forms of learning and progressively modifying and elaborating them as successive approximations to an adequate theory are confronted by new facts. . . . My preferred strategy was not to discard the original concept but rather to extend it to a broader conception of associations among representations of events, and in such a way that Stimulus-Response connections would be simply a special case of the more general concept, perhaps clearly exemplified only in some forms of conditioning for human beings and in learning of lower organisms. (p. 7)

Like Stimulus Sampling Theory, adaptive networks provide a framework for theory development that builds cumulatively on the associative concepts originally established for simpler forms of learning. By extending assumptions regarding the representation of events, and the nature of the stimulus-response connections to be modified, the distributed network model integrates the powerful learning rule from the Rescorla–Wagner conditioning model with the stimulus-representation assumptions from Stimulus Sampling Theory. Each of these models has had a long and successful history within learning theory, the former being applied primarily to animal learning data and the latter being most often associated with human learning phenomena. The preliminary explorations reported here suggest that, within the formalisms of adaptive network theory, some of the shortcomings of each of these earlier models might be addressed by the strengths of the other.

ACKNOWLEDGMENTS

For their thoughtful comments and advice on this work I am indebted to Gordon Bower, W. K. Estes, Michael Hee, Irene Laudeman, Douglas Medin, Robert Nosofsky, Richard Shiffrin, and Robert Solso. This research was supported by Grant N00014-83K-0238 from the Office of Naval Research.

REFERENCES

Anderson, J. A., Silverstein, J. W., Ritz, S. A., & Jones, R. S. (1977). Distinctive features, categorical perception, and probability learning: Some applications of a neural model. *Psychological Review, 84*, 413–451.

Atkinson, R. C. (1958). A Markov model for discrimination learning. *Psychometrika, 23*, 308–322.

Atkinson, R. C., Bower, G. H., & Crothers, E. J. (1965). *Introduction to mathematical learning theory*. New York: Wiley.

Atkinson, R. C., & Estes, W. K. (1963). Stimulus sampling theory. In R. D. Luce, R. R. Bush, & E. Galanter (Eds.), *Handbook of mathematical psychology*. New York: Wiley.

Binder, A., & Estes, W. K. (1966). Transfer of response in visual recognition situations as a function of frequency variables. *Psychological Monographs: General and Applied, 80*(23), 1–26.

Binder, A., & Feldman, S. E. (1960). The effects of experimentally controlled experience upon recognition responses. *Psychology Monograph, 74*(496).

Binder, A., & Taylor, D. (1969). Effects of frequency and novelty in transfer. *Journal of Experimental Psychology, 80*, 91–94.

Blough, D. S. (1975). Steady-state data and a quantitative model of operant generalization and discrimination. *Journal of Experimental Psychology: Animal Behavior Processes, 104*, 3–21.

Bower, G. H. (1972). Stimulus-sampling theory of encoding variability. In E. Martin & A. Melton (Eds.), *Coding theory and memory*. Washington, DC: V. H. Winston.

Bower, G. H., & Hilgard, E. R. (1981). *Theories of learning*. New Jersey: Prentice-Hall.

Bush, R. R., & Mosteller, F. (1955). *Stochastic models for learning*. New York: Wiley.

Elman, J. L., & Zipser, D. (1988). Learning the hidden structure of speech. *Journal of the Acoustical Society of America, 83*(4), 1615–1626.

Estes, W. K. (1955). Statistical theory of spontaneous recovery and regression. *Psychological Review, 62*, 145–154.

Estes, W. K. (1959). Component and pattern models with Markovian interpretation. In R. R. Bush & W. K. Estes (Eds.), *Studies in mathematical learning theory* (pp. 9–52). Stanford, CA: Stanford University Press.

Estes, W. K. (1964). Probability learning. In A. W. Melton (Ed.), *Categories of human learning*. New York: Academic Press.

Estes, W. K. (1982). *Models of learning, memory, and choice*. New York: Praeger.

Estes, W. K., Campbell, J. A., Hatsopoulos, N., & Hurwitz, J. B. (1989). Base-rate effects in category learning: A comparison of parallel network and memory storage-retrieval models. *Journal of Experimental Psychology: Learning, Memory, & Cognition, 15*(4), 556–571.

Estes, W. K., & Hopkins, B. L. (1961). Acquisition and transfer in pattern versus component discrimination learning. *Journal of Experimental Psychology, 61*, 322–328.

Feldman, J. A. (1985). Connectionist models and their applications: Introduction. *Special Issue of Cognitive Science, 9*, 1.

Flagg, S. F., & Medin, D. L. (1973). Constant irrelevant cues and stimulus generalization in monkeys. *Journal of Comparative and Physiological Psychology, 85*(2), 339–345.

Gluck, M. A. (1991). Stimulus generalization and representation in adaptive network models of category learning. *Psychological Science, 2*(1), 50–55.

Gluck, M. A., & Bower, G. H. (1986). Conditioning and categorization: Some common effects of informational variables in animal and human learning. In *Proceedings of the Eighth Annual Conference of the Cognitive Science Society,* Amherst, MA. Hillsdale, NJ: Lawrence Erlbaum Associates.

Gluck, M. A., & Bower, G. H. (1988a). From conditioning to category learning: An adaptive network model. *Journal of Experimental Psychology: General, 117*(3), 225–244.

Gluck, M. A., & Bower, G. H. (1988b). Evaluating an adaptive network model of human learning. *Journal of Memory and Language, 27,* 166–195.

Gluck, M. A., & Bower, G. H. (1990). Component and pattern information in adaptive networks. *Journal of Experimental Psychology: General, 119*(1), 105–109.

Gluck, M. A., Bower, G. H., & Hee, M. R. (1989). A configural-cue network model of animal and human associative learning. In *Proceedings of the Eleventh Annual Conference of the Cognitive Science Society.* Ann Arbor, MI. Hillsdale, NJ: Lawrence Erlbaum Associates.

Hanson, S. J. (1990). A stochastic version of the delta rule. *Physica D, 42,* 265–272.

Hilgard, E. R., & Bower, G. H. (1975). *Theories of learning.* New Jersey: Prentice-Hall.

Hinton, G. E., & Anderson, J. A. (1981). *Parallel models of associative memory.* Hillsdale, NJ: Lawrence Erlbaum Associates.

Hinton, G. E., McClelland, J. L., & Rumelhart, D. E. (1986). Distributed representations. In D. E. Rumelhart & J. L. McClelland (Eds.), *Parallel distributed processing: Explorations in the microstructure of cognition (Vol. 1: Foundations).* Cambridge, MA: Bradford Books/MIT Press.

Hull, C. L. (1943). *Principles of behavior.* New York: Appleton-Century-Crofts.

Kamin, L. J. (1969). Predictability, surprise, attention, and conditioning. In B. A. Campbell & R. M. Church (Eds.), *Punishment and aversive behavior* (pp. 279–296). New York: Appleton-Century-Crofts.

LaBerge, D. L. (1961). Generalization gradients in a discrimination situation. *Journal of Experimental Psychology, 62,* 88–94.

LaBerge, D. L. (1962). A recruitment theory of simple behavior. *Psychometrika, 27*(4), 375–396.

Lashley, K. S. (1929). *Brain mechanisms and intelligence.* Chicago: University of Chicago Press.

Levine, M. (1970). Human discrimination learning: The subset sampling assumption. *Psychological Bulletin, 74,* 397–404.

Lovejoy, E. (1968). *Attention in discrimination learning.* San Francisco: Holden-Day.

Mackintosh, N. J. (1975). A theory of attention: Variations in the associability of stimuli with reinforcement. *Psychological Review, 82,* 276–298.

McClelland, J. L., & Rumelhart, D. E. (1986). *Parallel distributed processing: Explorations in the microstructure of cognition (Vol. 2: Psychological and biological models).* Cambridge, MA: Bradford Books/MIT Press.

Medin, D. L. (1976). Theories of discrimination learning and learning set. In W. K. Estes (Ed.), *Handbook of learning and cognitive processes.* Hillsdale, NJ: Lawrence Erlbaum Associates.

Medin, D. L., & Edelson, S. M. (1988). Problem structure and the use of base rate information from experience. *Journal of Experimental Psychology: General, 117,* 68–85.

Medin, D. L., & Robbins, D. (1971). Effects of frequency on transfer performance after successive discrimination training. *Journal of Experimental Psychology, 87*(3), 434–436.

Neimark, E. D., & Estes, W. K. (1967). *Stimulus Sampling Theory.* San Francisco, CA: Holden-Day.

Parker, D. (1986). A comparison of algorithms for neuron-like cells. In *Proceedings of the Neural Networks for Computing Conference.* Snowbird, UT.

Pearce, J. M., & Hall, G. (1980). A model for Pavlovian learning: Variations in the effectiveness of conditioned and unconditioned stimuli. *Psychological Review, 87,* 532–552.

Rescorla, R. A. (1968). Probability of shock in the presence and absence of CS in fear conditioning. *Journal of Comparative and Physiological Psychology, 66,* 1–5.

Rescorla, R. A. (1976). Stimulus generalization: Some predictions from a model of Pavlovian conditioning. *Journal of Experimental Psychology: Animal Behavior Processes, 2,* 88–96.

Rescorla, R. A., & Wagner, A. R. (1972). A theory of Pavlovian conditioning: Variations in the effectiveness of reinforcement and nonreinforcement. In A. H. Black & W. F. Prokasy (Eds.), *Classical conditioning II: Current research and theory.* New York: Appleton-Century-Crofts.

Restle, F. (1957). Theory of selective learning with probability reinforcements. *Psychological Review, 64,* 182–191.

Robbins, D. (1970). Stimulus selection in human discrimination learning and transfer. *Journal of Experimental Psychology, 84,* 282–290.

Rudy, J. W., & Wagner, A. R. (1975). In W. K. Estes (Ed.), *Handbook of Learning and Memory, Vol. 2.* Hillsdale, NJ: Lawrence Erlbaum.

Rumelhart, D. E., & McClelland, J. L. (1986). *Parallel Distributed Processing: Explorations in the microstructure of cognition (Vol. 1: Foundations).* Cambridge, MA: MIT Press.

Sejnowski, T. J. (1988). Neural populations revealed. *Nature, 332,* 308.

Shanks, D. R. (1989). Connectionism and the learning of probabilistic concepts. *Quarterly Journal of Experimental Psychology.*

Spence, K. W. (1936). The nature of discrimination learning in animals. *Psychological Review, 43,* 427–449.

Stone, G. O. (1986). An analysis of the delta rule and the learning of statistical associations. In D. E. Rumelhart & J. L. McClelland (Eds.), *Parallel distributed processing: Explorations in the microstructure of cognition (Vol. 1: Foundations).* Cambridge, MA: Bradford Books/MIT Press.

Suppes, P., & Atkinson, R. C. (1960). *Markov learning models for multiperson interactions.* New York: Stanford University Press.

Sutherland, N. S., & Mackintosh, N. J. (1971). *Mechanisms of animal discrimination learning.* New York: Academic Press.

Sutton, R. S., & Barto, A. G. (1981). Toward a modern theory of adaptive networks: Expectation and prediction. *Psychological Review, 88,* 135–170.

Thompson, R. F. (1965). The neural basis of stimulus generalization. In D. J. Mostofsky (Ed.), *Stimulus generalization* (pp. 154–178). Stanford, CA: Stanford University Press.

Thorndike, E. L. (1898). Animal intelligence: An experimental study of the associative processes in animals. *Psychological Review, Monograph Supplement, 2*(8), 28–31.

Wagner, A. R. (1969). Stimulus selection and a modified continuity theory. In G. Bower, & J. Spence (Eds.), *The psychology of learning and motivation (Vol. 3.)* New York: Academic Press.

Wagner, A. R., & Rescorla, R. A. (1972). Inhibition in Pavlovian conditioning: Applications of a theory. In R. A. Boakes & S. Halliday (Eds.), *Inhibition and learning* (pp. 301–36). New York: Academic Press.

Widrow, B., & Hoff, M. E. (1960). Adaptive switching circuits. *Institute of Radio Engineers, Western Electronic Show and Convention, Convention Record, 4,* 96–194.

10 Serial Organization in a Distributed Memory Model

Bennet B. Murdock
University of Toronto

INTRODUCTION

In this chapter I extend the chaining model of Lewandowsky and Murdock (1989) to serial organization in general, and to chunking in particular. The chaining model is the serial-order part of TODAM, a Theory of Distributed Associative Memory. This theory is an attempt to provide a unified conceptual framework for human memory, much as the stimulus-sampling theory of Estes did for conditioning and learning (Estes, 1955). The Estes theory was a pioneering effort, and the chapters in this volume provide eloquent testimony to its fruitfulness. Although I have a different perspective, my goal is exactly the same; namely, to try to provide a single unified comprehensive explanation for a broad range of memory phenomena. This paper honors Estes more by emulating his general approach than by implementing or applying the specific details of his approach.

According to the chaining model, a serial string ABCDE is stored as a set of overlapping digrams (AB, BC, CD, DE) and retrieved by running off the string (A-B, B-C, C-D, and D-E). The items in the string are represented by random vectors and digrams are formed by convolving the two item vectors. Thus, associations are represented by convolutions. These convolutions are stored by summation in a common memory vector, and the retrieval operation is correlation. To run off the string, each successive item is correlated with the memory vector, and the retrieved information resembles the next item. The retrieved information must be deblurred for successful recall. The chaining model can account for much of the basic serial-order data in short-term memory, but it cannot account for serial organization or chunking in any simple or natural way.

We extend the model by introducing multiple convolutions, n-grams, and chunks. We start with a brief overview, beginning with the notion of multiple convolution. If the items in the string are A, B, C, . . . , then in the model these items would be represented by N-dimensional item vectors **a**, **b**, **c**, . . . The convolution operator is denoted by "∗". Thus, **a**∗**b** is the convolution of two item vectors, or a two-way convolution; **a**∗**b**∗**c** is a three-way convolution; **a**∗**b**∗**c**∗**d** is a four-way convolution, etc. Because **a**, **b**, **c**, **d**, . . . are vectors, their convolution is also a vector.[1]

The effect of grouping is to determine where breaks come. Thus, if we had ABC, DE, FGH, we would have three multiple convolutions (**a**∗**b**∗**c**, **d**∗**e**, and **f**∗**g**∗**h**). Because each multiple convolution is a single vector, it can function as a unit. These multiple convolutions **a**∗**b**∗**c**, **d**∗**e**, and **f**∗**g**∗**h** can function as units at a higher level even though they contain item vectors as constituents at a lower level.

We use the term <u>n-gram</u> for the n-way autoconvolution of the sum of n item vectors. An autoconvolution is the convolution of a vector with itself; for example, **a**∗**a** would be a two-way autoconvolution, or **a**∗**a**∗**a** would be a three-way autoconvolution. For DE, the 2-gram (or digram) would be

$$(\mathbf{d} + \mathbf{e}) * (\mathbf{d} + \mathbf{e}) = \mathbf{d} * \mathbf{d} + \mathbf{e} * \mathbf{e} + 2(\mathbf{d} * \mathbf{e}).$$

The three constituents (**d**∗**d**, **e**∗**e**, and **d**∗**e**) are vectors, so the n-gram, the sum of the three, is also a vector. For ABC, the 3-gram (or trigram) would be

$$(\mathbf{a} + \mathbf{b} + \mathbf{c}) * (\mathbf{a} + \mathbf{b} + \mathbf{c}) * (\mathbf{a} + \mathbf{b} + \mathbf{c}) = \mathbf{a} * \mathbf{a} * \mathbf{a} + \mathbf{b} * \mathbf{b} * \mathbf{b} + \mathbf{c} * \mathbf{c} * \mathbf{c} +$$
$$3(\mathbf{a} * \mathbf{a} * \mathbf{b} + \mathbf{a} * \mathbf{a} * \mathbf{c} + \mathbf{a} * \mathbf{b} * \mathbf{b} + \mathbf{b} * \mathbf{b} * \mathbf{c} + \mathbf{a} * \mathbf{c} * \mathbf{c} + \mathbf{b} * \mathbf{c} * \mathbf{c}) + 6(\mathbf{a} * \mathbf{b} * \mathbf{c}).$$

Thus, the n-gram includes the multiple convolution **a**∗**b**∗**c** but it also includes three multiple autoassociations (i.e., **a**∗**a**∗**a**, **b**∗**b**∗**b**, and **c**∗**c**∗**c**) and all possible cross-products (e.g., **a**∗**a**∗**b**, **a**∗**a**∗**c**). The n-gram makes recognition, recall, and redintegration all possible, given the right retrieval cues.

For serial organization, we need these retrieval cues; this is where the chunk comes in. A chunk is defined as the sum of all the n-grams that have been formed for a serially-organized string. Thus, if the string was ABC then the chunk would be

$$\mathbf{a} + (\mathbf{a} + \mathbf{b}) * (\mathbf{a} + \mathbf{b}) + (\mathbf{a} + \mathbf{b} + \mathbf{c}) * (\mathbf{a} + \mathbf{b} + \mathbf{c}) * (\mathbf{a} + \mathbf{b} + \mathbf{c}).$$

In this case, the chunk is the sum of three n-grams: an engram, a digram, and a trigram. For retrieval, the delta vector (a vector whose middle element is 1.0 with all other elements zero) correlated with the chunk retrieves **a**′, an approximation to the first item that must be deblurred to **a** for recall to be successful. Then **a**

---

[1]Technically speaking, for this to be true we must use an infinite-dimensional vector space with all but the center N elements set equal to zero.

correlated with the chunk retrieves **b'**, and **b'** must be deblurred to **b**. Finally **a∗b** correlated with the chunk retrieves **c'**, which must be deblurred to **c**.

A chunk includes several n-grams; more precisely, it includes as many n-grams as there are items in the string. In this example, the chunk includes 3 n-grams: the engram **a**, the digram **(a + b)∗(a + b)**, and the trigram **(a + b + c)∗(a + b + c)∗(a + b + c)**. Multiple convolutions are included in n-grams, and n-grams are included in chunks, so multiple convolutions are included in chunks.

We suggest that chunks are stored in memory, and that these chunks form the basis for serial organization. We have worked out some of the details of the chunking model, but as yet the applications to data are mostly qualitative. There are many empirical phenomena in the area we have to account for; we have only made a beginning. To indicate what these phenomena are, we turn next to a brief account of some of the more important data in the field.

SERIAL ORGANIZATION

Serial organization refers to the structuring of sequential information in memory. Some sort of hierarchical organization is generally suggested (Johnson, 1972), but the mechanisms involved are seldom specified (Estes, 1972). For greater generality these mechanisms should apply to grouping in short-term memory tasks and serial organization in learning, and should not be incompatible with more complex cognitive tasks such as discourse processing, reading, and reasoning.

Experimental Findings

The optimal group size for unrelated information in short-term memory seems to be three or four (Ryan, 1969; Wickelgren, 1964). Depending on the material, a hierarchical structure seems to be more appropriate than a "matrix" (independence) structure when order errors are analysed in terms of within- versus between-group position (McNicol & Heathcote, 1986). Pauses in presentation not only facilitate grouping but seem to provide the opportunity for some sort of chunking to occur (Wilkes, 1975). Error gradients for groups generally have the shape of an inverted U (Lee & Estes, 1977), and Transition Error Probabilities (TEPs) are generally highest at the group boundaries (Johnson, 1972). Much the same pattern occurs with interresponse time measures (McLean & Gregg, 1967). In serial learning, a very stable organization seems to develop over trials, but items between groups are almost never recalled (Martin & Noreen, 1974). For a more complete review see Crowder (1976) or Murdock (1974).

We can think of these groups as chunks (Miller, 1956); Johnson (1970, 1972) has suggested that a chunk is an "opaque container." To use a homely metaphor, items in a chunk are like objects in a suitcase. You must open up the suitcase to find out what objects are inside. Chunks at one level can combine to form a "superchunk" at a higher level, and a complex hierarchical structure can result from this pyramiding.

Reordering items other than the first does not greatly impair learning, but changing group structure from ABC-DEFG to, for example, AB-CDE-FG results in zero transfer (Johnson, 1972). Bower and Winzenz (1969) suggest a "reallocation hypothesis"—the first item of a chunk serves as its address, and changes to the first item are most disruptive. This, of course, is quite consistent with the TEP pattern reported by Johnson.

In his famous article, Miller (1956) suggested that memory span was 7 ± 2. If we consider memory span measured in items a measure of memory "capacity," and we consider each item a chunk, then memory capacity would be seven chunks. Mandler (1967) argued that seven was an overestimate; a more realistic estimate was closer to five. Broadbent (1975) argued that even five was too high; he suggested the "magic number" was three. Although he tested several possible structures for working memory that might give rise to the results, he did not specify what the mechanisms or processes might be.

Baddeley (1986) has done considerable research on working memory and suggested two components, a spatial-visual scratch pad and an articulatory-verbal loop. He has not, however, extended this analysis to grouping and organization. The Estes perturbation model, by contrast, was developed explicitly to deal with these phenomena (Estes, 1972). This model is a hierarchical model that assumes separate or localized storage of individual items or chunks. It is discussed more fully later.

Serial Recall

The notion of multiple convolutions has been applied to serial recall by Liepa (1977; see Murdock, 1979). Memory for a 5-item list would be represented as

$$\mathbf{M} = \mathbf{a} + \mathbf{a}*\mathbf{b} + \mathbf{a}*\mathbf{b}*\mathbf{c} + \mathbf{a}*\mathbf{b}*\mathbf{c}*\mathbf{d} + \mathbf{a}*\mathbf{b}*\mathbf{c}*\mathbf{d}*\mathbf{e}$$

where $\mathbf{a}$, $\mathbf{b}$, $\mathbf{c}$, $\mathbf{d}$, and $\mathbf{e}$ represent the separate item vectors and $\mathbf{M}$ is the memory vector. Then at recall,

$$\delta \# \mathbf{M} = \mathbf{a}' \rightarrow \mathbf{a},$$

$$\mathbf{a} \# \mathbf{M} = \delta + \mathbf{b}' \rightarrow \mathbf{b},$$

$$(\mathbf{a}*\mathbf{b}) \# \mathbf{M} = \delta + \mathbf{c}' \rightarrow \mathbf{c},$$

$$(\mathbf{a}*\mathbf{b}*\mathbf{c}) \# \mathbf{M} = \delta + \mathbf{d}' \rightarrow \mathbf{d}, \text{ and}$$

$$(\mathbf{a}*\mathbf{b}*\mathbf{c}*\mathbf{d}) \# \mathbf{M} = \delta + \mathbf{e}' \rightarrow \mathbf{e}.$$

The delta vector (δ) retrieves $\mathbf{a}'$ because $\mathbf{a}$ is the only engram in $\mathbf{M}$. After the first item, the retrieved information includes δ, but this can easily be filtered out.

There are many advantages to this approach to serial recall. Each probe or retrieval cue has a unique target (e.g., $(a*b*c)\#(a*b*c*d) = d'$ where **d** is the target). What has been retrieved to date functions as the probe for the next item. If the retrieved information is not correctly deblurred the situation is not necessarily catastrophic; the retrieved information can be used in the multiple convolution for the next probe. Little serial-to-paired-associate transfer would be expected (Young, 1968) because, with one exception (**a*b**), the stored convolutions are not digrams. (See Lewandowsky & Murdock, 1989, for a more detailed discussion of this point.) The Ranschburg effect (Wickelgren, 1965) is not a problem for the same reason. (The Ranschburg effect refers to the [surprisingly little] effect of repeating an item in the same list. If the list were ABCDBE then one has two choices after B, B-C or B-E.)

Recall of Missing Items

Yntema and Trask (1963) illustrated the recall of missing items with an example. Present a list of five words to a subject, pause, then present four of these five words in a scrambled order and ask for the recall of the missing item. In an informal experiment, they found about 80% recall of the missing item. They argued that it was hard to imagine how a person could do this task without making a search through the first list to test whether each item had appeared in the second list.

Multiple convolutions provide an alternative interpretation for the recall of missing items. Say the study list was ABCDE and the subset or probe list was BDEA. Then

$$(b*d*e*a)\#(a*b*c*d*e) = c'$$

and if the subject can deblur c' into **c** then he or she can report the missing item **c**. All you do is form a multiple convolution of the memory set, a multiple convolution of the probe set, correlate the latter with the former and out pops the missing item. Note that the order of the items in the probe set is immaterial because convolution is commutative; we show this more formally later.

Jung (1965) compared the recall of presented elements and the recall of missing elements (ROPE and ROME) of an exhaustive category. ROPE was only superior to ROME when the smallest number (13) of elements (the 50 U.S. states) was presented. When 25 or 37 elements were presented, ROPE and ROME did not differ statistically. For the latter case, the percentage of correct recalls was higher for ROPE than ROME, but the errors of commission were also higher.

Buschke introduced the "missing scan" technique, the fixed-set equivalent of Yntema and Trask's varied-set procedure and the special case of ROME when there is only a single missing element. Buschke (1963a) showed that performance dropped off as set size increased, as would be predicted because of the growth of variance with n. He also reported a clear serial position effect for recall of the missing item (Buschke, 1963b). The effect was a monotonic recency effect up to a one- or two-item primacy effect. His categories were sets of digits (e.g., 1–14; 4–16) and he found that measuring memory capacity this way gave a higher estimate than the more conventional memory-span technique. He argued that the missing scan technique gave a purer capacity measure than the conventional technique because it was not contaminated by output interference.

Bower and Bostrom (cited in Kimble, 1967, pp. 42–44) used a repetition procedure like Murdock and Babick (1961) to study repetition effects in the missing scan. In a free recall situation, Murdock and Babick repeated a preselected target item in list after list until it was recalled. The purpose of this manipulation was to see if repeated presentations of an item were independent. As far as could be determined, they were. In the missing-scan study, Bower and Bostrom obtained the same result. There was no benefit to repeating a missing item. (A fixed-set procedure was used, and the same item was omitted from different permutations of the remaining items until it was correctly reported.)

At a qualitative level, serial recall and recall of missing items have a natural interpretation in terms of multiple convolutions. The problem with Liepa's scheme for serial recall and using multiple convolutions for ROME is that multiple convolutions alone are not powerful enough to do all the serial tasks we know subjects can do. These include such obvious and important tasks as item recognition and probe recall. No item recognition would be possible as no item information has been stored, nor could the subject respond to either a sequential or a positional probe. Obviously something more is needed. The chunking model answers that need.

THE CHUNKING MODEL

Convolution

We start with the notion of convolution. A pairwise convolution, say $f*g$, is used to represent the association of two N-dimensional item vectors f and g. This view of association is very different from the traditional view where an association is seen as a connection or link between two discrete items or nodes. It is more like a holistic or Gestalt view, where the separate entities are completely intertwined. In a convolution, the components of f are dispersed or distributed over the components of g; conversely, the components of g are dispersed or distributed over the components of f.

The convolution operator is applied to item vectors that are assumed to be center-justified random vectors; that is, vectors whose elements are random variables centered on zero. As in all previous applications of TODAM, we assume that the elements of the random vectors are random samples from a normal distribution with mean zero and variance P/N where P is the power or the norm of the vector (Anderson, 1973), and N is the dimensionality of the space. In general we assume $P = 1$, but we address this point later in the section on Normalization.

At the component level, the convolution of two vectors **f** and **g** at the point of alignment x is defined as

$$(\mathbf{f}*\mathbf{g})\ (x) \equiv \Sigma_i f(i)g(x - i).$$

One way to visualize this summation is to think of the outer-product matrix of **f** and **g** and sum over the antidiagonals. Each antidiagonal represents a different value of x; x ranges from $-(N - 1)/2$ to $+ (N - 1)/2$, N odd.

Correlation

If convolution disperses the components of **f** and **g**, then correlation compacts them, or brings them back to their original form. The correlation of two vectors, say **f** and **h** at the point of alignment x, is defined as

$$(\mathbf{f}\#\mathbf{h})\ (x) \equiv \Sigma_i f(i)h(x + i)$$

In terms of the outer-product matrix, use the diagonals, not the antidiagonals; otherwise, everything else applies.

The possibility of using the convolution-correlation formalism for memory storage and retrieval was suggested by Borsellino and Poggio (1973). Convolution forms the mathematical basis for the holographic metaphor (Gabor, 1948; Pribram, 1971), though that is more for waveforms than for random vectors. The convolution-correlation formalism was applied by Liepa (1977; see Murdock, 1979) to the storage and retrieval of item, associative, and serial-order information. In TODAM it forms the basis for associative information (Murdock, 1982) and serial-order information (Lewandowsky & Murdock, 1989), and has also been used in the CHARM model of Eich (1982, 1985).

Convolution is commutative and associative so

$$\mathbf{f}*\mathbf{g} = \mathbf{g}*\mathbf{f} \text{ and}$$

$$\mathbf{f}*(\mathbf{g}*\mathbf{h}) = (\mathbf{f}*\mathbf{g})*\mathbf{h}.$$

Convolution is distributive over addition so

$$\mathbf{f}*(\mathbf{g} + \mathbf{h}) = (\mathbf{f}*\mathbf{g}) + (\mathbf{f}*\mathbf{h}).$$

The reason convolution and correlation mimic storage and retrieval is that, for single item vectors, if we let $\mathbf{h} = \mathbf{f}*\mathbf{g}$ then $\mathbf{f}\#\mathbf{h} = \mathbf{f}\#(\mathbf{f}*\mathbf{g}) = (\mathbf{f}\#\mathbf{f})*\mathbf{g} = \delta'*\mathbf{g} = \mathbf{g}'$ where $\mathbf{g}'$ resembles **g** (Eich, 1982). (δ' resembles δ, the delta vector, as long as the item vector **f** is "noiselike"; see Schonemann, 1987). For recall, $\mathbf{g}'$ must be deblurred to **g**; for recognition, one can use the inner (or dot) product of two vectors as the comparison operation (Anderson, 1973).

Multiple Convolution

We need some symbolism for multiple convolutions. For multiple autoconvolutions we use f^{*L} where

$$f^{*L} \equiv f*f*\ldots*f \text{ (L times); or}$$

$$f^{*1} \equiv f, \quad f^{*k} \equiv f*f^{*(k-1)}, \quad k > 1.$$

For multiple cross-convolutions we use the symbol X ("chi") where

$$X_{j=1}^{L}f_j \equiv f_1*f_2*\ldots*f_L$$

and $X_{j=1}^{1}f_j \equiv f_1$. In this context, X for convolution is like Σ (for addition) or Π (for multiplication).

The storage-retrieval formalism can be extended to multiple convolutions where the "item" vectors are themselves convolutions of item vectors. Thus,

$$(f*g)\#(f*g*h) = h'.$$

To show this, let $u \equiv f*g$. Then,

$$u\#(u*h) = (u\#u)*h = \delta'*h = h'$$

and because $u \equiv f*g$

$$(f*g)\#(f*g*h) = \{(f*g)\#(f*g)\}*h = \delta'*h = h'.$$

More generally, just by the proper choice of definitions, we can write

$$(X_{i=1}^{m}f_i)\#(X_{i=1}^{n}f_i) = X_{i=m+1}^{n}f_i', \quad n > m. \tag{1}$$

This works for any ordering of the item vectors within the two sets of parentheses, as convolution is commutative and associative.

This is a surprising result. If we take the case where $n = m + 1$ then we see that

$$(X_{i=1}^{m}f_i)\#(X_{1=1}^{m+1}f_i) = f_n'.$$

Thus, no matter how widely we disperse a single item vector, say f, over all the others (by multiple convolutions), we will be able to extract this same item vector f (or more properly f') by correlating their m-way convolution with the n-way convolution. Thus, correlation with multiple convolutions acts as a filter to retrieve a missing item.[2]

This shows how one can recall a missing item without any search or scan of the possible set. Correlate the convolution of the $n - 1$ items in the probe set with the convolution of the n items in the memory set and you retrieve the missing item. As previously noted, the order of the items in the subset is imma-

---

[2]Whereas the variance of the resemblance between f and f' increases with n, the expected value does not change.

terial because convolution is commutative. In the proof, the assignment of subscripts to items is arbitrary, so we can always identify the missing item as Item n.

N-grams and Chunks

If we use $G(n)$ for the n-gram of size n, then for $n = 3$ we have

$$G(n) = (a + b + c)^{*3} = (a + b + c)*(a + b + c)*(a + b + c).$$

A convolution expands just like a polynomial; for example,

$$(x + y + z)^3 = x^3 + y^3 + z^3 + 3(x^2y + x^2z + xy^2 + y^2z + xz^2 + yz^2) + 6xyz,$$

so expanding the terms we have

$$G(n) = a^{*3} + b^{*3} + c^{*3} + 3(a^{*2}*b + a^{*2}*c + a*b^{*2} + b^{*2}*c + a*c^{*2} + b*c^{*2}) + 6(a*b*c).$$

More generally we could write an n-gram $G(n)$ as

$$G(n) = (\Sigma_{i=1}^n f_i)^{*n}$$

where f_i are the n item vectors whose sum enters into the n-way autoconvolution.[3]

A chunk can then be defined as the sum of all n-grams 1 to n; thus, using the notation $C(n)$ to represent a chunk of size n then

$$C(n) = \Sigma_{j=1}^n G(j) = \Sigma_{j=1}^n (\Sigma_{i=1}^j f_i)^{*j}.$$

For the example where $n = 3$ the chunk $C(n)$ would be

$$C(n) = a + (a + b)^{*2} + (a + b + c)^{*3}$$

which, of course, could be expanded as shown.

There are a number of reasons for this definition of a chunk:

1. This definition embodies the notion of a chunk as an opaque container as the n-grams in the chunk, a vector of dimension $n(N - 1) + 1$, and the items within the n-grams are all combined into a single entity, and are "invisible" until the chunk is unpacked. You don't generally unpack a chunk into its constituents, but you could if you had to.

2. This definition implements the equivalent of a temporal-to-spatial mapping as the order of presentation is represented in the constituent n-grams: a, then $(a + b)^{*2}$, then $(a + b + c)^{*3}$, etc. In a sense the order information is represented by nesting; a is included in $a + b$, $a + b$ is included $a + b + c$, and so on.

---

[3]The idea of summing item vectors before associating them was first suggested to me in the context of a matrix model by Patrick Goebel; it comes originally from Cooper (1973).

However, storage and retrieval are intimately connected; at each step of the process, the retrieval operation must be performed with the right arguments for this "temporal-to-spatial" mapping to work.

3. In a way, the first item represents the "address" of the chunk as it is the only single item in the chunk. Thus, for any chunk size n

$$\delta\#C(n) = \delta\#\{\Sigma_{j=1}^n G(j)\} = \delta\#\{G(1) + G(2) + \ldots + G(n)\} \cong \delta\#G(1)$$
$$= G'(1) \cong f_1$$

where f_1 is the first item. That is, f_1 is the only single item in the chunk, and all the higher order terms that the delta vector extracts are noise. We can always show this by going to the component level because

$$E[Z(Z + ZW + ZWX + \ldots)] = E[Z^2 + Z^2W + Z^2WX + \ldots] =$$
$$E[Z^2] + E[Z^2W] + E[Z^2WX] + E[\ldots] = E[Z^2] = 1$$

where Z, W, and X are independent random variables and $E[Z^2W]$, $E[Z^2WX]$, $\ldots$, = 0.

4. A chunk can always be unpacked into its constituent units. Thus, if we use Liepa's scheme and omit the retrieved delta vectors for clarity we have

$$\delta\#C(n) = \delta\#G(1) = a' \rightarrow a$$
$$a\#C(n) = a\#G(2) = \alpha a' + \beta b' \rightarrow b,$$
$$(a*b)\#C(n) = (a*b)\#G(3) = \alpha a' + \beta b' + \gamma c' \rightarrow c, \text{ and so on}$$

At each step after the first the retrieved information always includes a linear combination of the target item and the items recalled (retrieved) to date. How this linear combination can be deblurred is discussed later.

Of course, deblurring will not always work, nor should it, as even short-term memory is not infallible. The probability of deblurring (and consequently the probability of recall) will decrease as the variance increases, and the variance will increase as n increases. Again this is as it should be, but we will need derivations like those in Weber (1988) before we know whether the quantitative details are correct.

5. Chunks can always be combined into "superchunks," and in principle one can form a hierarchy to any desired level. Thus, if we denote the j-th chunk of size n or $C_j(n)$ as

$$C_j(n) = \Sigma_{k=1}^n (\Sigma_{i=1}^k f_{ji})^{*k}$$

then the p-th "superchunk" of size n' or $C_p'(n')$ would be

$$C_p'(n') = \Sigma_{k'=1}^{n'}\{\Sigma_{i=1}^{k'} C_{pi}(n)\}^{*k'} = \Sigma_{k'=1}^{n'}\{\Sigma_{i=1}^{k'}(\Sigma_{1=1}^{i'} f_{k'i})^{*i'}\}^{*k'}.$$

This nesting can be continued to still higher levels, though of course chunks are "noisier" (more variance) than items, superchunks are noisier than chunks, and so on.[4] Any pattern of repetition of items between chunks (or chunks between superchunks) is possible.

6. At the lower level, chunks contain the necessary ingredients to perform most if not all of the standard short-term memory tasks: serial recall, item recognition, probe recall, probe recognition, and (of course) recall of missing items. A few parameters will have to be added to generate the myriad serial-position effects that exist (e.g., primacy and recency effects, item and order errors, similarity effects, grouping effects, memory span and running memory span, short-term forgetting, the buildup and release from proactive inhibition, repetition effects, learning and transfer) but we want to keep the formulation as general as possible for as long as possible.

7. Item information can be derived from a chunk; it does not have to be explicitly stored during the formation of a chunk. Given Equation 1 it is also the case that

$$\mathbf{F}^{*m} \# \mathbf{F}^{*n} = \mathbf{F}^{*(n-m)}$$

so if $m = n - 1$ then

$$\mathbf{F}^{*(n-1)} \# \mathbf{F}^{*n} = \mathbf{F}$$

and if

$$\mathbf{F} = \Sigma_{i=1}^{n} \mathbf{f}_i$$

then

$$(\Sigma_{i=1}^{n} \mathbf{f}_i)^{*(n-1)} \# (\Sigma_{i=1}^{n} \mathbf{f}_i)^{*n} = \Sigma_{i=1}^{n} \mathbf{f}_i$$

where $\Sigma_{i=1}^{n} \mathbf{f}_i$ is the desired item information; that is, $\mathbf{f}_1 + \mathbf{f}_2 + \ldots + \mathbf{f}_n$.[5]

Thus, in forming a chunk, if we have available the next-to-last multiple convolution (i.e., $\mathbf{F}^{*(n-1)}$) we can correlate the next-to-last multiple convolution with the last multiple convolution (i.e., $\mathbf{F}^{*n}$) and generate all the item information that went into $\mathbf{F}$ (i.e., $\mathbf{f}_1, \mathbf{f}_2, \ldots, \mathbf{f}_n$). This way we can have both item and associative information available for the memory vector. We return to this point when we discuss Working Memory.

Normalization

So far I have avoided the question of normalization; however, this is a fundamental issue. If you do not normalize the chunks then the power (or norm of the vector) will grow without limit. This growth poses two problems: First, in any

---

[4]Because of the variance, higher order chunks might need nice clean labels where we could label a chunk with $\mathbf{L}$ by forming $\mathbf{L}_* \mathbf{C}_i(n)$ or $\mathbf{C}_2\{\mathbf{L}, \mathbf{C}_i(n)\}$.

[5]It is also the case that, if $m = n$, then $\mathbf{F}^{*m} \# \mathbf{F}^{*n} = \mathbf{F}^{*n} \# \mathbf{F}^{*n} = \delta$, so $\mathbf{F}^{*0} = \delta$.

real system, no growth without limit can occur. Second, this growth can introduce a serious confounding into the comparison process. We use the dot (or inner) product to compare two vectors (e.g., the probe vector and the memory vector), and the result is assumed to be a direct reflection of encoding factors and forgetting. If the norm of a chunk is greater than the norm of an item vector then they are no longer commensurate. Or, more accurately, there is a confounding: The dot product will reflect encoding, forgetting, and a compounding effect; to assess encoding or forgetting factors one must somehow correct for the compounding (i.e., the effect of combining items into n-grams and n-grams into chunks). Such a correction seems antithetical to the notion of a chunk; it would be far better if we could bring items and chunks to the same scale (norm) so that they would be directly comparable.[6]

How to do this is not an easy matter. My own view is to distinguish between a normative approach and a functional approach. According to a normative approach, one should normalize n-grams and chunks to P the power of the item vectors (after Anderson, 1973), and in general P (or the norm of the vector) is set to 1. Alternatively, with a functional approach one can use parameters to achieve approximately the same end, though the approximation will be imperfect. Even though it might be quite bad locally, it could be relatively satisfactory globally.

The justification for distinguishing between a nominal and a functional approach is, I think, that a distributed memory model such as TODAM should be like the Theory of Signal Detection—a theory of the Ideal Observer/Memorizer. Thus, the nominal approach is prescriptive; it specifies how the memory system should normalize. The functional approach is not prescriptive—instead, it characterizes how the memory system does in fact normalize the chunk, at least approximately.

Consider an n-gram. An n-gram is the n-way convolution of the sum of n item vectors. So, to normalize the sum all we need is

$$G(n) = \left(\Sigma_{i=1}^{n} \frac{1}{n} \mathbf{f}_i \right)^{*n} = n^{-n}(\Sigma_{i=1}^{n} \mathbf{f}_i)^{*n}$$

which translates into $\mathbf{a}$, $.25(\mathbf{a} + \mathbf{b})^{*2}$, $.037(\mathbf{a} + \mathbf{b} + \mathbf{c})^{*3}$, and so on. Since the power of an n-way autoconvolution grows as P^n, if $P = 1$ then normalizing autoconvolutions is not a problem. To average the n-grams in a chunk we can use the same rule as for items; namely

$$C(n) = \Sigma_{i=1}^{n} \frac{1}{n} \left(\Sigma_{j=1}^{i} \frac{1}{i} \mathbf{f}_j \right)^{*i}$$

This then would be the normative approach. Functionally we could use

$$C(n) = \Sigma_{i=1}^{n} \beta_i (\Sigma_{j=1}^{i} \alpha_j \mathbf{f}_j)^{*i}$$

---

[6]The memory store can only be normalized with respect to one type of comparison. If the store contains, for example, engrams, digrams, and trigrams, then it looks very different depending upon whether the probe is a single item, a pair, or a triple.

where $\beta_i = \beta^{n-i}$ and $\alpha_j = \alpha^{i-j}$ so α normalizes the items within an n-gram and β normalizes the n-grams within a chunk.[7]

Terminology

Although items, n-grams, and chunks are all vectors, they are vectors at different levels. They are also different sorts of vectors as the component distributions are quite different. An n-gram is the result of the convolution of the sum of item vectors, and a chunk is the sum of n-grams. For convenience, let us term the sum of n-item vectors a "mixture," and the normalized sum a "compound."

To distinguish among levels, let us use $\mathbf{f}$ as the general symbol for an item vector. At the component level

$$\mathbf{f} = (\ldots, f(-1), f(0), f(1), \ldots)$$

where there are N non-zero terms (the components). Because $\mathbf{f}$ is a random vector, the components of $\mathbf{f}$ are simply N random variables.

Let us call the sum of n item vectors a mixture and denote it by $\mathbf{F}$; thus

$$\mathbf{F} = \Sigma_{i=1}^n \mathbf{f}_i.$$

We will denote a compound, a normalized mixture by $\bar{\mathbf{F}}$. For the normative case we would have

$$\bar{\mathbf{F}} = \Sigma_{i=1}^n \frac{1}{n} \mathbf{f}_i$$

and for the functional case we would have

$$\bar{\mathbf{F}} = \Sigma_{i=1}^n \alpha_i \mathbf{f}_i$$

where generally $\alpha_i = \alpha^{n-i}$. Thus, as noted, there are two ways of normalizing a mixture to form a compound. Regardless, we will denote a compound by $\bar{\mathbf{F}}$ where $\mathbf{F}$ is a mixture.

An n-gram, denoted $\mathbf{G}(n)$, is the n-way autoconvolution of a compound. Thus

$$\mathbf{G}(n) = (\bar{\mathbf{F}})^{*n} = \left(\Sigma_{i=1}^n \frac{1}{n} \mathbf{f}_i \right)^{*n}$$

in the normative case. This immediately suggests a possible reason why the optimal group size in studies of short-term memory is so small (on the order of 3 or 4). The item information in an n-gram is proportional to n^{-n}, and because this decreases rapidly with n $(1, \frac{1}{4}, \frac{1}{27}, \frac{1}{250}, \ldots)$ the item information becomes attenuated so rapidly that it soon becomes virtually useless.

A chunk $\mathbf{C}(n)$ is the sum of n-grams 1 to n; thus, in the functional case

$$\mathbf{C}(n) = \Sigma_{p=1}^n \{\mathbf{G}(n)\} = \Sigma_{p=1}^n (\Sigma \bar{\mathbf{F}})^{*p} = \Sigma_{p=1}^n \beta_p (\Sigma_{\ell=1}^p \alpha_\ell \mathbf{f}_\ell)^{*p}$$

---

[7]α_i and β_i are just shorthand to denote that α (or β) is a function of i; if we write the difference equation $\mathbf{M}_j = \alpha\mathbf{M}_{j-1} + \mathbf{f}_j$, then at the end of list presentation $\mathbf{M}(L) = \Sigma_{i=i}^L \alpha_i \mathbf{f}_i$ where $\alpha_i = \alpha^{L-i}$.

We do not distinguish between a normalized and an unnormalized chunk—all chunks should be normalized. A chunk is still a vector even though it is composed of the sum of n-grams where an n-gram is the autoconvolution of a compound, which in turn is a normalized mixture, which finally is the sum of n item vectors whose components are n random variables.

We probably have labeled chunks. After all, a chunk is an opaque container and it probably has a name, so the name would be its label. We need labels to retrieve chunks. The delta vector will retrieve the first item of a chunk, but that is when we want to unpack it. For higher order cognitive processes we want the chunk to function as a unit; that is why labels are probably necessary.

Working Memory

It is assumed that the operations (convolution and correlation) necessary to form a chunk go on in a Working Memory. This was called the "CPU" in Murdock (1982, p. 623, Fig. 7) but subsequently I changed it to "Working Memory" (Murdock, 1983, p. 321, Fig. 2). The properties of Working Memory have been extensively investigated by Baddeley and his associates (Baddeley, 1986), but I am concerned here with showing how Working Memory could be used to implement the operations required in the chunking model.

In the formation and utilization of chunks, multiple convolutions must be formed at storage and retrieval. To show how this can be done, we assume that Working Memory consists of a fixed number of arrays for vector processing. Even though registers or array processors may seem incompatible with the idea of a distributed memory model, the model requires operations on vectors and these operations must be performed somewhere (see also Mewhort, 1990). If these operations are carried out in Working Memory, and if they require arrays, then it would seem we have to assume Working Memory contains array processors.

To put some constraints on the model, let us assume that Working Memory consists of five array processors. This estimate is consistent with the buffer model of Atkinson and Shriffrin (1968) and with SAM (Gillund & Shiffrin, 1984), and is also consistent with some of the early data on short-term memory capacity limitations. Even though this estimate may be much too small (see Schneider & Detweiler, 1988), it would be nice to show that a very limited-capacity Working Memory could do the job.

To show how the multiple convolutions can be formed and used, denote the five arrays of Working Memory as v, w, x, y, and z. A flow chart to illustrate how a chunk could be formed is shown in Fig. 10.1.

The inner loop forms the n-gram, and the middle loop sums the engram to form the chunk. The outer loop repeats for the next chunk. Probabilistic encoding would be done in the P system (see Murdock, 1982) and normalization could be done in w. There is also some bookkeeping involved, but that is not shown. The test for i is discussed in the next section.[8]

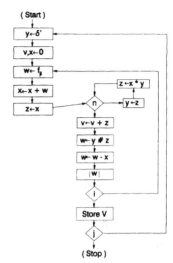

FIG. 10.1. Flow chart to show one could use a Working Memory with five array processors (v, w, x, y, and z) to sum n-grams to form a chunk. Here, f_{ji} is the i-th item in the j-th chunk.

A comparable flow chart for retrieval is shown in Fig. 10.2. Here you only need four array processors (w, x, y, and z). The initialization with the delta vector is necessary to retrieve the first item, and then you loop to retrieve the 2nd, 3rd, . . . , nth item. An approximation to the delta vector can be obtained by correlating any normalized vector with itself; for example,

$$\delta' = f \# f$$

These flow charts make three points. First, the necessary operations can be performed within the constraints of a rather restricted Working Memory. Second, there is no need for the item vectors to be stored individually either during storage or retrieval; they may be pooled as the process develops. Third, multiple convolutions can be used as probe items during retrieval almost as simply as single items; all it requires are the two extra steps $z \rightarrow y$ and $x*y \rightarrow z$.

Stop Rule

The chunk, involving as it does the sum of multiple autoconvolutions, is a form of associative information. As noted, one can derive the item information from the associative information. Again, as in Equation 1,

$$F^{*(n-1)} \# F^{*n} = F$$

---

[8]You could eliminate the v array processor by storing the n-grams directly and not forming explicit chunks. This would restrict normalization possibilities and would also change the notion of a chunk considerably.

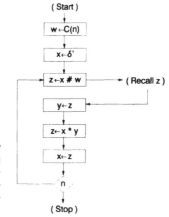

(Start)

w ←C(n)

x←δ'

z←x # w ──────→ (Recall z)

y←z

z←x * y

x← z

n

(Stop)

FIG. 10.2. Flow chart to show how one could use a Working Memory with four array processors (w, x, y, and z) to retrieve the n items of a short serial string.

and, as $F = \Sigma_{i=1}^{n} f_i$, we can retrieve the item information simply by correlating the next-highest with the highest-order convolution. This also works if we use $\bar{F}$ for F as we should, though now we will get the normalized sum of the item vectors. All we need, then, is an extra step whereby we include F (or $\bar{F}$) in memory as well as $C(n)$.

As shown in Fig. 10.1, each n-gram loop ends with $F^{*(n-1)}$ in Register y and F^{*n} in Register z. Thus, all we have to do is correlate the contents of Register y with the contents of Register z and we immediately have the desired item information. This provides a possible "stop rule" in the formation of a chunk. We say a chunk is n items, but how big is n? Or in Fig. 10.1, what terminates the middle loop?

Suppose we compare the derived item information with the direct item information. The direct item information, or $\bar{F}$, always exists in Register x, so if we difference x and w (which hold y#z) we will have the difference between the direct and the derived item information. If the difference is Δ and we have an error criterion ϵ then our stop rule would be of the form "if $\Delta < \epsilon$, continue; otherwise stop" where we define the absolute difference Δ as

$$\Delta = |\Sigma_{i=1}^{n} f_i - (F^{*(n-1)} \# F^{*n})|;$$

that is, the absolute difference between the encoded item information and the derived item information.[9]

Because the variance of F^{*n} increases with n so will the variance of $F^{*(n-1)} \# F^{*n}$. As a result, Δ will increase with n and so $\Delta > \epsilon$ becomes progressively more likely as n increases. If optimum grouping occurs in the 3–4

---

[9]You could also use the dot product for the $\Delta < \epsilon$ test.

range, then this could be the point at which Δ typically exceeds ϵ. Of course, for a fixed ϵ, n would covary with N, the dimensionality of the space, so it would be nice to have some studies of grouping with a wide variety of different stimulus material.

Deblurring

Chunks are more powerful than n-grams, and n-grams are more powerful than multiple convolutions. By "more powerful" I mean that they endow the memory system with the capability of remembering more and retrieving more. They are more powerful because chunks subsume n-grams and n-grams subsume multiple convolutions. However, there is a price to pay: not only is the variance higher, but in recall one must deblur a linear combination of responses.

To illustrate, first assume we only stored multiple convolutions as in Liepa's CADAM scheme. Then

$$\mathbf{M} = \mathbf{a} + \mathbf{a}_*\mathbf{b} + \mathbf{a}_*\mathbf{b}_*\mathbf{c}$$

and at recall

$\delta\#\mathbf{M} = \mathbf{a'} \rightarrow \mathbf{a},$

$\mathbf{a}\#\mathbf{M} = \delta + \mathbf{b'} \rightarrow \mathbf{b},$ and

$(\mathbf{a}_*\mathbf{b})\#\mathbf{M} = \delta + \mathbf{c'} \rightarrow \mathbf{c}$

Now suppose instead that we stored the chunk composed of these n-grams; namely (and ignoring normalization)

$$\mathbf{M} = \mathbf{a} + (\mathbf{a} + \mathbf{b})^{*2} + (\mathbf{a} + \mathbf{b} + \mathbf{c})^{*3}.$$

At recall we would have

$\delta\#\mathbf{M} = \mathbf{a'} \rightarrow \mathbf{a},$

$\mathbf{a}\#\mathbf{M} = \delta + \alpha\mathbf{a'} + \beta\mathbf{b'} \rightarrow \mathbf{b},$ and

$(\mathbf{a}_*\mathbf{b})\#\mathbf{M} = k\delta + \alpha\mathbf{a'} + \beta\mathbf{b'} + \gamma\mathbf{c'} \rightarrow \mathbf{c}$

where α, β and γ are the coefficients of the linear combination. Even though the target item will always be the strongest item—see the polynomial expansion shown—we still have to explain how a linear combination of responses can be deblurred. (See Jordan, 1986, for the basic concepts here.)

Again assume we can use item information for this purpose. Modify the chunk retrieval scheme as follow:

$\delta\#\mathbf{M} = \mathbf{a'} \rightarrow \mathbf{a},$

$(\mathbf{a}\#\mathbf{M}) - \alpha\mathbf{a} = \delta + \alpha\mathbf{a'} + \beta\mathbf{b'} - \alpha\mathbf{a} \cong \delta + \beta\mathbf{b'} \rightarrow \mathbf{b},$

$(\mathbf{a}_*\mathbf{b})\#\mathbf{M} - (\alpha\mathbf{a} + \beta\mathbf{b}) = \delta + \alpha\mathbf{a'} + \beta\mathbf{b'} + \gamma\mathbf{c'} - \alpha\mathbf{a} - \beta\mathbf{b} \cong \delta + \gamma\mathbf{c'} \rightarrow$
$\mathbf{c}$

or more generally

$$(X_{i=1}^{k-1}f_i)\#\mathbf{M} - \Sigma_{i=1}^{k-1}f_i \cong \delta + \mathbf{f}_k' \to \mathbf{f}_k.$$

That is, we subtract out the item information retrieved to date from the retrieved information, and are left with (an approximation to) the target item.

There is room in Working Memory to do this; we can accumulate the item information in a separate array as we go along. There is a normalization problem: How do we set the α's and β's at output to match the α's and β's at input? There is not space to discuss this issue here, but it may not be a major problem. Conditional-probability analyses show that the probability of recall of Item j is greater if Item j-1 is recalled than if it is not recalled (McDowd & Madigan, 1991; Morton, 1970; Murdock, 1976). The subtraction method should be less effective if the retrieved information for Item j-1 were substituted for the target Item j-1.

Is there any evidence that "noise" can be filtered out by simple subtraction? In fact there is, at least in the perceptual systems. In audition, the threshold for a weak signal in noise played to one ear is reduced if the identical noise (but not the signal) is also played in the other ear (see Durlach & Colburn, 1978, for a review). A comparable effect to "binaural unmasking" has recently been discovered in vision (Schneider, Moraglia, & Jepson, 1989). The auditory and visual systems could do this by filtering out the noise by simple subtraction; if the auditory and visual systems have this capability, so could the memory system.

Obviously we have not solved the deblurring problem. One still has to deblur a noisy version of the target item into the target item itself. But perhaps it is fair to suggest that this filtering method turns deblurring from a major problem to a minor problem.

Item and Order Errors

Item and order errors have played a large role in theorizing about serial organization (e.g., Crowder, 1979; Healy, 1978; Murdock, 1987). Different rates of loss for item and order information has been used by Estes (1972) as a basic motivation for his perturbation model, and he places considerable weight on the shape of the distance functions obtained in his experiments.[10]

How does the chunking model explain item and order errors? It is assumed that successive chunks are stored in a common memory vector. Consequently, what is retrieved is a linear combination not only of items in the correct n-gram but also of items in the previous n-grams of the same size, decremented by a forgetting parameter. Thus, item errors could come from prior chunks of the same size and should show both recency and serial-position specificity. Intralist

---

[10]The distance function is the function showing the presentation position i for an item recalled in serial position j.

intrusions are a frequent type of item error (Conrad, 1965), and they show both these effects. Failure to deblur the retrieved information should lead to omission, another common type of item error.

How could order errors occur? Suppose that at the time of study there was a delay in the processing and by mistake the third item was encoded before the second item. Then the chunk would be

$$a + (a + c)*^2 + (a + c + b)*^3$$

and the expected recall order would be ACB. It is still an order error, but it occurred at encoding, not during storage.[11]

Another way order errors could occur is if the items were encoded in the right order but the number of autoconvolutions got out of phase. Suppose it lagged by one and we had

$$a + (a + b)*^1 + (a + b + c)*^2 + (a + b + c + d)*^3.$$

Then, if A is recalled, the next item recalled could well be C (we can't subtract out b because it hasn't been recalled) and then B and D are the two candidates for the next recall. We could have ACB . . . or ACDB or even ACD . . . B, so remote transpositions are not out of the question either.[12]

Thus, at least in principle, the chunking model can explain item and order errors, even though there is no discrete storage of separate items. Also, the model does not use the item-to-position associations used by some of the other models of serial-order effects. The contrast with the Estes perturbation model is particularly striking, and we close with a comparison of the two models so the reader will have a better appreciation of the differences between them.

COMPARISON WITH THE PERTURBATION MODEL

The fundamental difference between the perturbation model of Estes and his collaborators (Estes, 1972; Lee & Estes, 1977, 1981) and the chunking model presented here is in the storage of order information. In the perturbation model items have position tags—each item is directly associated to a specific position. There are no item-to-item associations. Within a string, we have a temporal-to-position mapping. After presentation there is a periodic reactivation ("rehearsal") and on each reactivation there is a small probability (θ) that an item will move one slot ahead or behind. (End items can only move inward.) There is also some probability ($\underline{a}$) that an item will lose its position tag (become free floating, as it were).

---

[11]This would not explain why transpositions tend to increase with the duration of the retention interval.

[12]Transpositions are most frequent between adjacent items and become progressively less frequent as the separation increases.

With these assumptions, one can set up a transition matrix and compute the probability distribution for the recall of any item in any position. These are the distance functions already mentioned, and for four-item strings the outside two items (1 and 4) are essentially mirror images of each other, as are the inner two items (2 and 3). All are single-peaked functions; the outside two tail inward while the middle two fall off on each side of the center. These functions become less sharp as the retention interval increases (see also Nairne, 1990), indicating a progressive loss of order information.

To account for loss of item information, a multi-level schema is envisaged with a hierarchy of trials, segments or groups within trials, and items within groups. Reactivation (rehearsal) goes on concurrently at all three levels, with attendant diffusion of order information at all three levels. There are two ways forgetting can occur. Perturbation at the trial level means that items from the wrong trial will replace items from the tested trial. The other way is the possibility that after a perturbation at the group level the item is not recoded in any position (see Equations A2–A5 in Lee & Estes, 1981). There is one parameter (θ) for perturbation and another parameter ($\underline{a}$), which is the probability that an item is recoded in a new position following a perturbation.

Perturbation goes on at all three levels, thus, we need three parameters (θ, θ_1, θ_2) for these three levels. Also, some decisions have to be made concerning the number of reactivations at each level at each retention interval studied; in the worst case, this would add 3t parameters where t is the number of different retention intervals used, but this could be reduced to three by controlling the timing of events in the experiment. There is the parameter $\underline{a}$ and the guessing probabilities, but in some cases the latter could be determined from the type of material used. In the best case, then, one would need seven parameters to fit the model to data or to derive predictions.

In the chunking model, by contrast, there are no position tags, items are combined into n-grams and n-grams into chunks, and time of arrival is preserved in the n-grams. The ith item makes its first appearance in the ith n-gram, and is absent in all prior n-grams. The chunk is an opaque container; you cannot tell in what position a given item occurred until you unpack the chunk.

Structure or organization is achieved very simply; chunks at a lower level become items at a higher level. There is no restriction on the repetition of items within or across chunks, or the number of levels one can have. When we use a chunk as an item at one level, we generally don't unpack it into its constituents at a lower level, but we could if we had to.

The perturbation model has been extended by the work of Healy and her colleagues. In a Hebb repetition experiment Cunningham, Healy, and Williams (1984) found that retention of retested lists and precued segments was too high, and they had to reduce θ and θ_2 by the same amount to account for the data. They suggested that this might be one way short-term activity traces could be frozen or crystallized into a longer-term structural memory trace. This modification implies that memory traces will deteriorate over time even with the cyclic reactivation, but this deterioration can be reduced by active rehearsal.

In a follow-up study, Healy, Fendrich, Cunningham, and Till (1987) extended the retention interval to distinguish between two possible explanations for the prior results. One was the change in the rehearsal process as previously described; the other was an initial encoding into long-term memory. With some probability $(1 - \alpha)$, an item is not subject to the perturbation process, and this turned out to be the better explanation. In terms of the theoretical analysis this seems to be almost a consolidation-like process where the long-term memory representation is strengthened over the retention interval.

In the chunking model, there is a distinction between Working Memory and a long-term store (M), but Working Memory is required to carry out the necessary storage and retrieval operations. Certainly rehearsal is possible, though that has not been explicitly discussed here. We can place sharp constraints on the size of Working Memory, whereas the perturbation model seems to require a short-term or episodic memory that can maintain a rather large number of items simultaneously.

SUMMARY

I have tried to extend the chaining model of Lewandowsky and Murdock (1989) to account for serial organization. The basic concepts are multiple convolutions, n-grams, and chunks. A chunk is defined as the sum of n-grams 1 to n where an n-gram is the n-way autoconvolution of the sum of n item vectors. As in previous applications of TODAM, items are represented as random vectors and stored in a common distributed memory, and correlation is used to retrieve items associated by convolution.

Why am I suggesting such a complex extension of TODAM? Basically it is because simpler extensions just won't do. When one goes beyond simple item and associative information, the data on serial-order effects show that the human memory system is very flexible in storing and retrieving serial-order information. We need a powerful model if we are to have any chance of understanding the breadth and depth of these effects.

Why do we need multiple convolution? Pairwise convolutions won't work for the recall of missing items. Suppose the study list (M) was ABCD and the probe (P) was CAD then

$$P \# M = (c*a + a*d)\#(a*b + b*c + c*d) = \text{Noise}$$

but

$$(c*a*d)\#(a*b*c*d) = b'$$

just as

$$a\#(a*b) = (a\#a)*b = \delta'*b = b'.$$

Why sum item vectors to form an n-gram? If you do, then a single n-gram is powerful enough to support recognition, recall, and redintegration given the right retrieval cues. Why autoconvolve this sum n times? Because as argued by Humphreys, Bain, and Pike (1989), pairwise associations are not enough,[13] and you can't recall missing items without the n-way cross-association. Why does one need to sum the n-grams to form a chunk? To have the right retrieval cues available for serial recall.

For purposes of recognition the item information can be derived from the chunk if the highest-order compound (the normalized sum of n item vectors) is available. In recall, the retrieval cue is the multiple convolution of all prior items, and deblurring is necessary to map the retrieved information into the target information. Even though the retrieved information is a linear combination of all prior items, these prior items can be differenced out if a representation of the items retrieved to date still exists. All these operations can be implemented in a Working Memory consisting of four or five array processors.

This composite storage of item and associative information is rich enough to support recognition, recall, and redintegration. It can easily do many simple episodic serial tasks such as item recognition, ordered recall, running memory span, and recall of missing items; it results in chunks that are opaque containers and qualitatively seems consistent with many grouping effects reported in the literature. In principle it can be extended upwards to provide hierarchical structure to any desired degree of complexity. However, most of the quantitative details have not been worked out, so how good the fit is of the model to the data remains to be seen.

ACKNOWLEDGMENTS

This work was supported by NSERC Grant APA146 from the Natural Sciences and Engineering Research Council of Canada. I would like to thank Stephan Lewandowsky, David G. Mitchell, and Richard Shiffrin for many helpful comments on the manuscript. Also, I have Dave Mitchell to thank for Footnotes 4, 6, 8, and 9. A program library of routines written in C he wrote for the chunking model is available on request.

---

[13]They argue that pairwise associations are not enough to explain the results from a cross-associate paradigm. In this paradigm, you can respond appropriately to a "semantic" or an "episodic" probe when the semantic pairs are strong associates like KING-QUEEN and BREAD-BUTTER, and the episodic pairs are newly-learned KING-BUTTER and BREAD-QUEEN cross-associates.

REFERENCES

Anderson, J. A. (1973). A theory for the recognition of items from short memorized lists. *Psychological Review, 80,* 417–438.

Atkinson, R. C., & Shiffrin, R. M. (1968). Human memory: A proposed system and its control processes. In K. W. Spence & J. T. Spence (Eds.), *The psychology of learning and motivation: Advances in research and theory, Vol. 2,* (pp. 89–195). New York: Academic Press.

Baddeley, A. D. (1986). *Working memory.* Oxford: Oxford University Press.

Borsellino, A., & Poggio, T. (1973). Convolution and correlation algebras. *Kybernetik, 122,* 113–122.

Bower, G. H., & Winzenz, D. (1969). Group structure, coding, and memory for digit series. *Journal of Experimental Psychology Monograph Supplement, 80, Part 2,* 1–17.

Broadbent, D. E. (1975). The magic number seven after fifteen years. In A. Kennedy & A. Wilkes (Eds.), *Studies in long term memory* (pp. 3–18). London: Wiley.

Buschke, H. (1963a). Retention in immediate memory estimated without retrieval. *Science, 140,* 56–57.

Buschke, H. (1963b). Relative retention in immediate memory determined by the missing scan method. *Nature, 200,* 1129–1130.

Conrad, R. (1965). Order error in immediate recall of sequences. *Journal of Verbal Learning and Verbal Behavior, 4,* 161–169.

Cooper, L. N. (1973). A possible organization of animal memory and learning. In B. Lundquist & S. Lundquist (Eds.), *Proceedings of the Nobel Symposium on Collective Properties of Physical Systems* (pp. 252–264). New York: Academic Press.

Crowder, R. G. (1976). *Principles of learning and memory.* Hillsdale, NJ: Lawrence Erlbaum Associates.

Crowder, R. G. (1979). Similarity and order in memory. In G. H. Bower (Ed.), *The psychology of learning and motivation: Advances in research and theory, Vol. 13* (pp. 319–353). New York: Academic Press.

Cunningham, T. F., Healy, A. F., & Williams, D. M. (1984). Effects of repetition on short-term retention of order information. *Journal of Experimental Psychology: Learning, Memory, and Cognition, 10,* 575–597.

Durlach, N. I., & Colburn, H. S. (1978). Binaural phenomena. In E. C. Carterette & M. P. Friedman (Eds.), *Handbook of perception, Vol. 4, Hearing* (pp. 365–466). New York: Academic Press.

Eich, J. M. (1982). A composite holographic associative recall model. *Psychological Review, 89,* 627–661.

Eich, J. M. (1985). Levels of processing, encoding specificity, elaboration, and CHARM. *Psychological Review, 92,* 1–38.

Estes, W. K. (1955). Statistical theory of spontaneous recovery and regression. *Psychological Review, 62,* 145–154.

Estes, W. K. (1972). An associative basis for coding and organization in memory. In A. W. Melton & E. Martin (Eds.), *Coding processes in human memory* (pp. 161–190). Washington, DC: Winston.

Gabor, D. (1948). A new microscopic principle. *Nature, 161,* 177.

Gillund, G., & Shiffrin, R. M. (1984). A retrieval model for both recognition and recall. *Psychological Review, 91,* 1–67.

Healy, A. F. (1978). A Markov model for the short-term retention of spatial location information. *Journal of Verbal Learning and Verbal Behavior, 17,* 295–308.

Healy, A. F., Fendrich, D. W., Cunningham, T. F., & Till, R. E. (1987). Effects of cueing on short-term retention of order information. *Journal of Experimental Psychology: Learning, Memory, and Cognition, 13,* 413–425.

Humphreys, M. S., Bain, J. D., & Pike, R. (1989). Different ways to cue a coherent memory system: A theory for episodic, semantic, and procedural tasks. *Psychological Review, 96,* 208–233.

Johnson, N. F. (1970). The role of chunking and organization in the process of recall. In G. H. Bower (Ed.), *The psychology of learning and motivation: Advances in research and theory, Vol. 4* (pp. 171–247). New York: Academic Press.

Johnson, N. F. (1972). Organization and the concept of a memory code. In A. W. Melton & E. Martin (Eds.), *Coding processes in human memory,* (pp. 125–159). Washington, DC: Winston.

Jordan, M. I. (1986). An introduction to linear algebra in parallel distributed processing. In D. E. Rumelhart & J. L. McClelland (Eds.), *Parallel distributed processing, Vol. 1* (pp. 365–422). Cambridge, MA: MIT Press.

Jung, J. (1965). Recall of presented and missing elements of an exhaustive category as a function of percentage of elements presented. *Proceedings of the Convention of the American Psychological Association, 1,* 55–56.

Kimble, D. P. (Ed.). (1967). *Learning, remembering, and forgetting, Vol. 2.* New York: New York Academy of Sciences.

Lee, C. L., & Estes, W. K. (1977). Order and position in primary memory for letter strings. *Journal of Verbal Learning and Verbal Behavior, 16,* 395–418.

Lee, C. L., & Estes, W. K. (1981). Item and order information in short-term memory: Evidence for multilevel perturbation processes. *Journal of Experimental Psychology: Human Learning and Memory, 7,* 149–169.

Lewandowsky, S., & Murdock, B. B. (1989). Memory for serial order. *Psychological Review, 96,* 25–57.

Liepa, P. (1977). Models of content addressable distributed associative memory (CADAM). Unpublished manuscript, University of Toronto.

Mandler, G. (1967). Organization and memory. In K. W. Spence & J. T. Spence (Eds.), *The psychology of learning and motivation: Advances in research and theory, Vol. 1* (pp 327–372). New York: Academic Press.

Martin, E., & Noreen, D. L. (1974). Serial learning: Identification of subjective subsequences. *Cognitive Psychology, 6,* 421–435.

McDowd, J., & Madigan, S. (1991). Ineffectiveness of visual distinctiveness in enhancing immediate recall. *Memory & Cognition, 19,* 371–377.

McLean, R. S., & Gregg, L. W. (1967). Effects of induced chunking on temporal aspects of serial recitation. *Journal of Experimental Psychology, 74,* 455–459.

McNicol, D., & Heathcote, A. (1986). Representation of order information: An analysis of grouping effects in short-term memory. *Journal of Experimental Psychology: General, 115,* 76–95.

Mewhort, D. J. K. (1990). Alice in wonderland or Psychology among the information sciences. *Psychological Research, 52,* 158–162.

Miller, G. A. (1956). The magical number seven, plus or minus two: Some limits on our capacity for processing information. *Psychological Review, 63,* 81–96.

Morton, J. (1970). A functional model for memory. In D. A. Norman (Ed.), *Models of human memory* (pp. 203–260). New York: Academic Press.

Murdock, B. B. (1974). *Human memory: Theory and data.* Potomac, MD: Lawrence Erlbaum Associates.

Murdock, B. B. (1976). Item and order information in short-term serial memory. *Journal of Experimental Psychology: General, 105,* 191–216.

Murdock, B. B. (1979). Convolution and correlation in perception and memory. In L. G. Nilsson (Ed.), *Perspectives in memory research: Essays in honor of Uppsala University's 500th Anniversary* (pp. 105–119). Hillsdale, NJ: Lawrence Erlbaum Associates.

Murdock, B. B. (1982). A theory for the storage and retrieval of item and associative information. *Psychological Review, 89,* 609–626.

Murdock, B. B. (1983). A distributed memory model for serial-order information. *Psychological Review, 90,* 316–338.

Murdock, B. B. (1987). Serial-order effects in a distributed-memory model. In D. S. Gorfein & R. R. Hoffman (Eds.), *Memory and learning: The Ebbinghaus centennial conference* (pp. 277–310). Hillsdale, NJ: Lawrence Erlbaum & Associates.

Murdock, B. B., & Babick, A. J. (1961). The effect of repetition on the retention of individual words. *American Journal of Psychology, 74,* 596–601.

Nairne, J. S. (1990). Similarity and long-term memory for order. *Journal of Memory and Language, 29,* 733–746.

Pribram, K. H. (1971). *Languages of the brain.* Englewood Cliffs, NJ: Prentice-Hall.

Ryan, J. (1969). Grouping and short-term memory: Different means and patterns of grouping. *Quarterly Journal of Experimental Psychology, 21,* 137–147.

Schneider, B., Moraglia, G., & Jepson, A. (1989). Binocular unmasking: Analogue to binaural unmasking? *Science, 243,* 1479–1481.

Schneider, W., & Detweiler, M. (1988). A connectionist control architecture for working memory. In G. H. Bower (Ed.), *The psychology of learning and motivation, Vol. 21* (pp. 54–119). New York: Academic Press.

Schonemann, P. H. (1987). Some algebraic relations between involutions, convolutions, and correlations with applications to holographic memories. *Biological Cybernetiks, 56,* 367–374.

Weber, E. U. (1988). Expectation and variance of item resemblance distributions in a convolution-correlation model of distributed memory. *Journal of Mathematical Psychology, 32,* 1–43.

Wickelgren, W. A. (1964). Size of rehearsal group and short-term memory. *Journal of Experimental Psychology, 68,* 413–419.

Wickelgren, W. A. (1965). Short-term memory for repeated and non-repeated items. *Quarterly Journal of Experimental Psychology, 17,* 14–25.

Wilkes, A. L. (1975). Encoding processes and pausing behavior. In A. Kennedy & A. Wilkes (Eds.), *Studies in long term memory* (pp. 19–42). London: Wiley.

Yntema, D. B., & Trask, F. P. (1963). Recall as a search process. *Journal of Verbal Learning and Verbal Behavior, 2,* 65–74.

Young, R. K. (1968). Serial learning. In T. R. Dixon and D. L. Horton (Eds.), *Verbal behavior and general behavior theory* (pp. 122–148). Englewood Cliffs, NJ: Prentice-Hall.

11 Reducing Interference in Distributed Memories Through Episodic Gating

Steven A. Sloman
David E. Rumelhart
Stanford University

A number of computational theories of human memory that appear in the current literature share three essential characteristics (e.g., Anderson, Silverstein, Ritz, & Jones, 1977; Eich, 1982; Kohonen, 1988; McClelland & Rumelhart, 1985; Murdock, 1982). First, they view memory as being composed of a large number of relatively simple units operating in parallel. Second, they retain information by distributing the representation of each to-be-remembered item over a large number of these units such that each unit can be involved in the representation of many items. Third, each theory proposes a scheme for the learning of representations through modification of unit values, or of connections between units, during item presentation. The major advantage of these theories is their correspondence to certain properties of human concept formation. Through experience with individual instances, they naturally form abstract categories based on family resemblance and generalize these categories to new instances.

As several critics and theorists have pointed out (Hinton & Plaut, 1987; McCloskey & Cohen, in press; Ratcliff, 1989; Sutton, 1986), parallel and distributed theories of memory are often unable to explain certain phenomena under conditions of sequential learning. The learning algorithms associated with these theories are generally constructed to find a set of unit values that can represent items that are shown many times to the networks in a mixed order, whereby each item is presented many times at intervals distributed across the set of items shown. However, when all the presentations of each item are massed, when learning is sequential, learning of later items will disrupt memory for earlier items. Whereas experiments on human memory would lead us to expect some of this kind of retroactive interference, simulations of these theories exhibit far too much interference. According to McCloskey and Cohen, these simulations show "catastrophic interference."

Catastrophic Interference

Interference is an immediate consequence of the characteristics just mentioned. Because the networks modify their units to accommodate each new item that is presented, and because items share representational units, then as later items are presented and units are modified to accommodate them, the units will no longer necessarily encode a solution for earlier items. The values of the units after the network has converged to a solution are correct for the item last shown, and not necessarily correct for less recent items. When item presentation is mixed, it is possible for the networks to be pulled back and forth between the various items' solutions until they find a solution that accommodates all the items at the same time. Back-propagation, one commonly used learning algorithm, was derived under the assumption that exemplars are randomly selected according to a fixed distributed (Rumelhart, Hinton, & Williams, 1986).

Using simulations, Lewandowsky and Murdock (in press) show that a version of Murdock's (1982) convolution-correlation model does not suffer from catastrophic interference. We consider the problem of catastrophic interference in detail only for the class of nonlinear parallel distributed processing models that incorporate back-propagation as their learning algorithm. This class of models is described fully in Rumelhart and McClelland (1986). To be consistent with their usage, the term "weight" is used to refer to a modifiable connection between units through which one unit can activate another. It is in the weights that information is stored. Units themselves take on temporary activation values that represent the current state of the network, and not historical information.

Bounds on Catastrophic Interference.   To observe catastrophic interference, the output of a memory retrieval operation is compared to the corresponding target output after the network has been presented with interfering material. Hinton and Plaut (1987) demonstrated that some retention can be found by using a more sensitive measure. In spite of disruption of the network's ability to reconstruct items, it can show savings in relearning (cf. Hetherington & Seidenberg, 1989). Savings will occur when an algorithm looks for a solution for new items in a region close to the solution of earlier items, as back-propagation does if a small enough learning rate is used. It is this partial retention that will allow a network to hone in on a full solution when presentation is nonsequential.

The majority of networks that have been studied use a "linear" activation rule in which the activation of a unit at each time step is equal to a linear combination of the activations of units to which it is connected. Most learning algorithms will not cause interference in such a network if vectors of activation input to *every layer* are mutually orthogonal. This can easily be seen by considering a simple network consisting of only two input units each connected to a single output unit. We associate the input "1 0" with output "0" and the orthogonal input "0 1" with "1." The net input to the output unit will be $1*w_1 + 0*w_2$ for the first input, where w_1 and w_2 are the weights out of the first and second input units respectively; thus the value of w_2 makes no difference to the output of the network when pattern 1 is presented. Similarly, for pattern 2, the value of w_1 makes no difference. Moreover, when learning the second input pattern using any standard learning rule, only w_2 will change. So learning pattern 2 will not affect the network's response to pattern 1. Learning of orthogonal vectors produces no interference irrespective of the dimensionality of the network. And if vectors are almost orthogonal, then interference will be small.

If input patterns were chosen at random, then orthogonality could play an important role in minimizing interference. If unit activations are symmetric around zero and input vectors have a sufficiently large dimensionality, then if unit activations have an equal probability of being selected, the expected dot product of any two input vectors would be zero (i.e., they would be orthogonal). Furthermore, networks of linear units set up to minimize sum squared error tend to maximize the mutual independence of units within each hidden layer (cf. Baldi & Hornik, 1989). This behavior, which may be generalizable to other networks, would act to maximize the orthogonality of hidden unit vectors. However, assuming that inputs are orthogonal to each other in general seems arbitrary. In most cases, some structure is expected in the input set, and this can be used to reduce dimensionality and make predictions. The assumption of mutual orthogonality places too great a constraint on how we represent inputs. So orthogonality is not sufficient to prevent catastrophic interference.

Brousse and Smolensky (1989) have shown that if inputs are members of a category that is structured such that strings are "regular expressions"; that is, inputs are composed of nonrandom concatenations of components, then interference is not a problem. On the contrary, items presented later in a sequence will often *enhance* memory for earlier items from the same category. When new items come from the same category as earlier items, learning trials with the new items has the effect of giving the network more experience with the regular structure that new and old items share. Therefore, the regular structure is better learned and performance on those earlier items improves. This is a plausible solution to the catastrophic interference problem in domains in which a regular structure could easily be incorporated into the input representation, such as many natural categories (e.g., biological taxonomies), speech, and language generally. However, humans and other animals can also remember and discriminate among arbitrarily composed episodes, such as lists composed of words, sounds, and pictures, that are experienced only once. Humans have memory for events that cannot be described with a regular structure. So regularity is not sufficient to deal with catastrophic interference.

In summary, several considerations encourage us to conclude that catastrophic interference is not a catastrophic problem for distributed memory theory. Undoubtedly, a variety of network architectural options exist that would further minimize the problem. And we may not want to minimize it at all. After all, retroactive interference is a pervasive phenomenon in the study of human memory, and a psychological model must exhibit somewhat. Also, the extent to which human learning can be described as sequential exposure to material is open to question (cf. Hetherington & Seidenberg, 1989). When learning arithmetic, for example, a child is never exposed to all problems that use a certain number, and then all problems that use another number. Rather, the child is exposed to a variety of problems, usually in a sensible order in which later learning builds on earlier learning. Nevertheless, despite all these considerations, we believe that the human ability to remember all sorts of different things under all sorts of presentation conditions, and most especially the ability to remember episodes composed of arbitrary objects and events that occur only once, merits a special kind of network architecture that will further reduce interference. We propose two architectural constraints. The first reduces interference within a context. The second reduces interference between contexts. The two constraints are first described conceptually. A final section describes our computer implementation along with several demonstrations.

Reducing Interference Within a Context—Willshaw's Network

We assume that individual items are sparsely distributed in the space of possible representations. Vectors are made relatively far from each other, with the result that they interfere with each other less. This is accomplished by assuming that m, the number of active features in the representation of any particular item, is much less than n, the total number of features available.

Consider the simple linear network and learning algorithm suggested by Willshaw (1981) and pictured in Fig. 11.1. Willshaw's network consists of n_1 binary input units and n_2 binary output units, all taking on values of 0 or 1. Each output unit is connected to every input unit by a weight of either 0 or 1. Output unit activations, a_i, are updated according to the following threshold function:

$$
a_i = \begin{cases} 1, & \text{if } \sum_{j=1}^{n_1} w_{ij} a_j \geq m_1. \\[2em] 0, & \text{if } \sum_{j=1}^{n_1} w_{ij} a_j < m_1. \end{cases}
\tag{1}
$$

in which j spans the set of input nodes, w_{ij} is 1 if a connection exists between input i and output j and 0 otherwise, and m_1 is the number of input nodes with

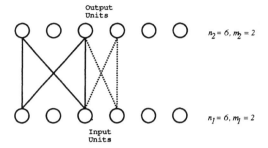

FIG. 11.1. Willshaw network with two items encoded auto-asso-ciatively. The item encoded by dotted lines is (0,0,1,1,0,0). The item encoded by solid lines is (1,0,1,0,0,0). An output unit is activated if and only if it is connected to all the activated input units.

value 1. In other words, an output unit is activated whenever there exists a connection between it and every activated input unit.

Willshaw suggested a simple learning scheme. To learn to activate a target output upon presentation of an input pattern, a connection must be created (the relevant weight set to 1) between each activated input unit (those with value 1) and each activated output unit. Learning a single input-output pair, therefore, consists of setting $m_1 m_2$ weights to 1. Each pair need only be presented once. Note that this scheme immediately eliminates half of the catastrophic inter-ference problem. As weights are only added, and never removed, old input-output pairs are never lost from storage. Problems in this network can only arise if too many such pairs are encoded. The kind of error that Willshaw's network will make, in proportion to the number of items learned, is to activate an output unit that was not associated with the input string.

To implement a memory model, we consider only the special case of auto-association, in which the desired output is equal to the input. In this case, $n_1 = n_2 = n$ and $m_1 = m_2 = m$. If we assume the m (out of n possible) activated input units are randomly chosen, then we can derive the following expression for the lowerbound of the capacity of this network in terms of r, the number of patterns that the network can learn before at least one spurious output bit is ensured:

$$r \leq .69(n/m)^2.$$

This expression is derived in the Appendix for the case in which m is constant across all patterns. Note that r increases with the square of system size (n), and decreases with the square of the size of each item (m). As an illustration, if $n = 10,000$ and $m = 50$, the system could learn 27,600 patterns before the proba-bility was one of an incorrect output bit being activated for any single pattern. Note that the analysis holds even if patterns are learned in a purely sequential order. Of course, in the following simulations that we report, we use much lower values of *m* and *n* to save computer time.

Reducing Interference Between Contexts

Tulving (1972) pointed out that the memory tasks used in the majority of studies of human memory were not a representative sample of the kinds of tasks performed by human memory. Up to that point, experimenters relied mostly on tasks that asked subjects to recall, identify, or discriminate the source of items or events; and when and where they had been exposed to them. He called these tasks "episodic" memory tasks because they tapped the ability to remember the contents of a particular episode. This kind of task can be distinguished from those tasks that simply ask subjects to do something with some material, usually to supply a correct answer, regardless of memory for the experience of being shown the material. Free association, lexical decision, and anagram completion are all examples of this class of nonepisodic memory tasks. Later, Tulving suggested that there is a memory system, a set of structures or processes or both, devoted to performance of episodic memory tasks, that is, devoted to the remembering of past experiences. He called this the *episodic memory system.*

As a means of reducing interference, we implement an episodic memory. Our network contains two kinds of information (cf. Humphreys, Bain, & Pike, 1989, for a similar move). One kind corresponds to the set of feature-feature relations encoded by the Willshaw network. This is the network's general knowledge base. A second kind acts as a gatekeeper, determining which set of relations will be operative in any given context. Thus, we make general knowledge contextually specific. How episodic units can gate the weights encoding general knowledge is pictured in Fig. 11.2. The knowledge base is modeled as a

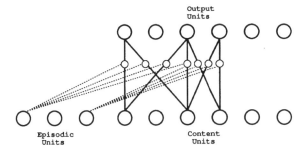

FIG. 11.2. Willshaw network with episodic gating. The leftmost episodic unit must be activated to allow activation to pass through any of the four leftmost Willshaw weights. The third episodic unit must be on to enable the relations encoded by the four rightmost connections. Representations of episodic contexts could also be distributed.

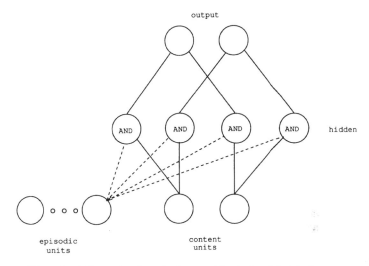

FIG. 11.3. Schematic of method to avoid catastrophic interference. All possible content to output pairings are present. One or more episodic units choose the relevant pairs in each context.

Willshaw network (solid lines). We refer to the input of this part of the network as "content" because it corresponds to the perceived content of an episode. The content units and episodic units together comprise the input units. Connected to each general knowledge weight by dotted lines are gatekeepers, the episodic units. Each general knowledge weight is active only if it is connected to all the active episodic units. The activation of these units represents an internally generated context that specifies a point in subjective time, place, and perhaps other internal parameters such as goals.

Figure 11.3 uses a localist representation to show the logic of the network with episodic units. Each general knowledge weight is associated with a dedicated hidden unit. This hidden unit acts as the junction between the general knowledge weight and the episodic weights. The solid lines in the figure represent the general knowledge weights that, in this case, include all possible connections between content features and output features. For instance, the rightmost solid lines of the upper and lower layers of weights correspond to the Willshaw weight connecting the rightmost content feature to the rightmost output feature. The network has the potential to instantiate any or all of the four possible connections between content and output units. We use an episodic unit, connected to the hidden units by dashed lines, to establish which of the content to output relations will be operative. If we conceive of the hidden units as AND gates, 1 being TRUE and 0 FALSE, then a 1 on a weight branching out of the episodic unit will instantiate its corresponding content to output connection and a 0 will render it inoperative. Other episodic units with different weights would be present to allow many different relations between content and output units, each operative in a different context. The same logic of episodic gating holds if representations in any part or all of the network are distributed.

By employing episodic gatekeepers, we are giving the system a way of compartmentalizing its knowledge so that any interference between items will occur mostly within a context. However, we imagine that contexts have distributed representations and a resultant similarity structure, just as we suppose items of knowledge do. Therefore, transfer and interference should occur between contexts in proportion to the similarity between those contexts, with no transfer and no interference obtaining whenever contexts are orthogonal.

Because knowledge is acquired through experience, learning always takes place within a context. Much of learning consists of binding old pieces of knowledge together in a new context. In terms of our theory, new learning will sometimes involve, especially in its early stages, the creation of new connections between content features and output features along with the construction of appropriate episodic connections. (We describe how the episodic weights are learned when we describe our implementation later.) More often though, especially after many general knowledge connections are in place, new learning consists of binding old links between content and output units to new contexts. For instance, to simulate an experiment that uses meaningful material (or material that subjects treat as meaningful), we use two phases of learning. Phase 1 models preexperimental learning. Many different items, some of which overlap with experimental items, are taught to the network using auto-association so that the network is able to reconstruct the content part of the input. The experiment itself is modeled in the second phase of learning. During this phase, a representation of the experimental context is used as input on the episodic unit. This could either remain constant for each episode, as in the simulation we report in the following, or change slowly (Mensink & Raaijmakers, 1988, demonstrate the explanatory power of a contextual fluctuation process in conjunction with the search of associative memory theory). Experimental items are taught (auto-associatively) in context. Because many (and sometimes all) of the general knowledge connections are learned in phase one, phase two learning consists mainly of the creation of new connections from the episodic units to the hidden units that gate the general knowledge weights.

Psychological Background

Various principles and phenomena of human memory are consistent with our use of an episodic memory to deal with catastrophic interference. The first is that in human memory, as in our system, memory depends on prior knowledge. Since work by Bartlett (1932), Bransford and Johnson (1972), Hyde and Jenkins

(1969), and Craik and Lockhart (1972), the pivotal role played by prior knowledge in acquiring new memories and in retrieving old ones has become uncontroversial. In our system, new memories are formed through novel combinations of old pieces of knowledge.

Our use of episodic units is also consistent with results obtained in the study of amnesia. The amnesic syndrome is characterized by a difficulty in performing episodic memory tasks in remembering the source of experienced events. But amnesics often do as well as normals at retrieving words and their meanings, at demonstrating motor skills, and at performing other nonepisodic memory tasks. Amnesics who perform very poorly on tasks of recognition and recall can exhibit normal memory facilitation (or "priming" effects) when tested by lexical decision (Moscovitch, 1982), homophone spelling (Jacoby & Witherspoon, 1982), stem and fragment completion (Graf, Squire, & Mandler, 1984; Warrington & Weiskrantz, 1974), or free association (Schacter, 1985). None of the latter tasks require subjects to identify the episode during which items were presented. Several theorists have argued that amnesia is a selective disruption of episodic memory (Cermak, 1984; Parkin, 1986; Schacter & Tulving, 1982; for contrary views see Moscovitch, 1982; Ostergaard, 1987; Squire & Cohen, 1984), though terminological disputes abound (cf. Mishkin, Malamut, & Bachevalier, 1984, for an identical conceptual distinction made between "memories" and "habits"). Because episodic memory can be selectively impaired, any model of memory should provide a distinct status for it.

New learning in our system occurs mostly in the episodic weights. Therefore, it is the information stored in those weights, the relations between earlier contexts and their contents, that we expect to suffer most from later learning. The relations between content features and output features will remain stable. Congruent with this hypothesis, experimental evidence puts the greater part of retroactive interference effects in episodic memory. Evidence comes from the study of forgetting and of the effect of different kinds of interpolated material on memory performance. Some changes of general knowledge weights will occur, and so we expect some long-term effects of retroactive interference on general knowledge. Some aspects of childhood amnesia could result from just such changes.

Some experiments have demonstrated strikingly little forgetting when memory is measured by tasks that do not tap episodic memory. Tulving, Schacter, and Stark (1982) showed subjects a list of 96 words and tested half the words an hour later and half the words one week later using word fragment completion (fill in the blanks, _J_ _D). Another group of subjects was tested using yes/no recognition. Fragment completion does not require that subjects remember the presentation episode whereas recognition does. The results are shown in Table 11.1. Fragments of words seen before (printed words) were completed more often than unprimed words, and the difference was about the same after one hour and after one week. But testing by episodic recognition indicated that some forgetting had taken place over the week; the hit rate was lower and the false alarm rate higher.

TABLE 11.1
Tulving, Schacter, and Stark (1982)

Fragment Completion	1 Hour	1 Week
primed	.47	.46
unprimed	.30	.32
Recognition		
hit rate	.78	.58
FA rate	.23	.33

Sloman, Hayman, Ohta, Law, and Tulving (1988) report facilitation of primed words more than 16 months after their presentation.

In the study of human memory, retroactive interference refers not to the detrimental effect of interpolating material between study and test, but to the relatively more detrimental effect of interpolating material similar to to-be-remembered material over material that is less similar. In the standard experiment, a list of words is presented one or more times. In the interference condition, the list is followed by another list of words. In the control condition, subjects are generally asked to do some nonlinguistic task, such as arithmetic. When memory is measured by free recall, performance is much lower in the interference condition (e.g., Tulving & Thornton, 1959). Interference effects have also been observed using list discrimination, that is, when subjects' task is to identify in which list an item appeared (Anderson & Bower, 1972; Deffenbacher, Carr, & Leu, 1981). Both these tasks depend on source retrieval on recollecting the context of learning. We posit that the cause of retroactive interference effects on these tasks is the disruption of source retrieval due to new spurious associations between items and contexts through the presentation of similar items in similar contexts. Similar vectors are those that share many common features and have few distinctive features. The interpolated context is bound to be similar to the original one. Therefore, some overlap between episodic features involved in original and interpolated learning is to be expected. If original and interpolated items are also similar, then we should also expect overlap in the general knowledge features involved in the two learning periods. Whenever interpolated learning causes original episodic features to gate pathways between original content features and interpolated output features, the likelihood of spurious activations during testing of the original list is increased. We attribute the effects of retroactive interference just described to such spurious activations.

A second kind of retroactive interference experiment uses pairs of items and manipulates the number of responses associated with each stimulus term. In the interference condition of this paradigm, each stimulus is shown with two or more responses (A-B, A-C, ...). In the control condition, each stimulus has a unique response (A-B, D-C, ...). The traditional task of cued recall (present A and test recall of B) inevitably shows appreciable retroactive interference (e.g., Barnes & Underwood, 1959, although they refer to their retroactive effects as "unlearning" because they asked for all responses to a given stimulus). Matching tasks, in which subjects are required to correctly pair up sets of stimuli and responses, can also exhibit associative interference, especially when a large number of lists are interpolated (e.g., Garskof & Sandak, 1964; Graf & Schacter, 1987). Retroactive effects on associative matching are not always observed, however, even when they are seen in cued recall (Postman & Stark, 1969). Nevertheless, as suggested by our model of paired-associate learning (described later), we suppose that when A has been associated with several responses, it is less predictive of any one, with the consequence of associative interference.

COMPUTER IMPLEMENTATION

Several auxiliary assumptions have been made to implement a computer model of a Willshaw network with episodic units. The implementation is pictured in Fig. 11.4 in which n content units are connected to n output units through n^2 hidden units. The pathways from content to output units are designed to retain all the properties of Willshaw's network. Hidden units serve only as gates, allowing episodic units to open and close pathways from content to output. Each content unit is connected to a single row of hidden units and each output unit to a single column. Thus there is a single path from each content unit to each output unit. This path corresponds to a single weight in Willshaw's scheme. Only weights between content and hidden units are variable, weights from hidden to output units are fixed at 1. The e episodic units are each connected to every hidden unit by a variable weight. We include many more episodic weights, en^2, than learnable general knowledge weights, of which there are only n^2, in accordance with our belief that new memories consist mostly of new episodic links between old pieces of knowledge.

The hidden units are designed to behave like the AND gates pictured in Fig. 11.3: Episodic connections multiplicatively gate the Willshaw-type general knowledge connections. The activation of hidden unit ij (which connects content unit j to output i) is a multiplicative function of the number of active episodic units that have established connections to the hidden unit, the activation of content unit j (a_j), and the weight connecting unit j to the hidden unit (w_{ij}):

$$h_{ij} = f(episodic\ netinput_{ij}) \cdot a_j \cdot f(w_{ij}),$$

where $a_j \cdot f(w_{ij})$ corresponds to a pathway in a Willshaw network.

$$episodic\ netinput_{ij} = \sum_{k=1}^{e} min(W_{ij,k};\ 1)a_k - \theta,$$

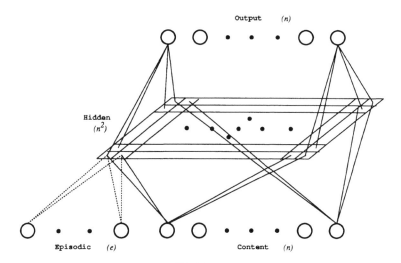

FIG. 11.4.   Implementation of Willshaw network with episodic gating.
Each output unit is connected by a fixed weight to a column of hidden
units. Each content unit is connected by a variable weight to a row of
hidden units so that a single path connects each content and output
unit (solid lines). Each episodic unit is connected to every hidden unit
by a variable weight (dotted lines). A single linear unit (not shown)
counts the number of active content units and subtracts the total from
the netinput to the output units.

in which k ranges over the set of episodic units, $W_{ij,k}$ is the weight from episodic
unit k to hidden unit ij, and a_k is the activation of episodic unit k (equal to either 0
or 1), and

$$\theta = m_{epis} - .5,$$

where m_{epis} is the number of active episodic units. Because we multiply each a_k
by a value no greater than one (due to the min function), *episodic netinput$_{ij}$* never
exceeds $m_{epis} - \theta = .5$. The function f is a logistic mapping real numbers into
the $[0,1]$ interval:

$$f(x) = 1/(1 + e^{-G*x}), \tag{2}$$

where G is a gain parameter (1/temperature). When testing the performance of
the system, G is set to a high value (such as 40) so that the hidden units come
close to acting as threshold units, as required by Willshaw's network. When G is
high, values greater than 0 are mapped to numbers close to 1, those less than 0 to
0. G is slowly increased during learning, for reasons discussed later.

    Because we apply the logistic to both *episodic netinput$_{ij}$* and w_{ij}, we are
assured that all three terms in the activation function of h_{ij} take on values
between 0 and 1; and when G is high, the values become essentially 0 *or* 1. In

this case, all three terms must be close to 1 to activate the hidden unit. This is achieved if and only if weights exist with values of near one or greater from every active episodic unit and from the one content unit connected to it that also must be active. Thus, each hidden unit mimics an AND gate in the sense that its activation is greater than zero only if it has established connections to all active episodic units and to an active content unit.

Output units are logistic units with constant gains of 4, designed to approximate the threshold units described in Equation 1. All weights projecting to them from the hidden units are fixed at one so that the netinput to output i is

$$netinput_i = \sum_{j=1}^{n} h_{ij} - (m - .5),$$

where m, as before, is the number of active content units. When G is high the h_{ij} are essentially 0 or 1 so that $netinput_i$ achieves a value of greater than 0 only if the number of units active in the ith column of hidden units is greater than or equal to the number of active content units. Therefore, only if episodic units have opened pathways between an output unit and every active content unit does that output unit become active.

Learning in this system uses a modified back-propagation procedure to mimic Willshaw's original learning algorithm. The learning rate is set to a high value (1.0) because, in Willshaw's network, weights were either added at their maximum value (1) or they weren't changed. Our algorithm has two new features. First, only positive errors are back-propagated. The learning algorithm thus never reduces a weight (weights can be reduced through weight decay).

Second, the gain on the hidden units, G in Equation 2, is started at a low value (about 1) and slowly increased over the course of learning up to values around 30. The gain schedule allows us to achieve two apparently conflicting objectives. On one hand, the gain should be as high as possible as we are trying to model Willshaw's threshold units that correspond to an infinite gain. On the other hand, we would like a low gain to speed up learning. In back-propagation, each weight increments is proportional to the derivative with respect to the netinput of the receiving unit's activation. Figure 11.5A displays the activation function of a logistic with a high gain. For all values of the function except those near .5, the derivative is clearly near zero. When the gain is lowered, as in Fig. 11.5B, we see that the slope is now well above zero for a much wider range of values, and thus learning is speeding up considerably. Using our gain schedule, learning is rapid early on, and we obtain the advantages of Willshaw's threshold units later.

DEMONSTRATIONS

The following demonstrations are intended to show some of the properties of our system, especially the absence of catastrophic interference. In all cases, $e = 20$,

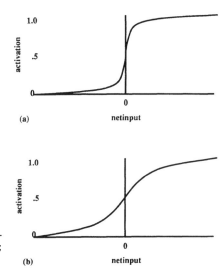

FIG. 11.5. (a) Logistic activation function with high gain; and (b) low gain.

$m_{epis} = 3$, $n = 50$, and $m = 4$. Gain was initially set at 1.04 and gradually increased (roughly exponentially) to 36.4 over 10 cycles. In most demonstrations, the error measure reported is total best match. This is the number of times that the actual output was closer to the desired output than to any other target in the test set, where distance was measured by sum squared error.

Single List Learning

To verify the capacity of the Willshaw network, we taught it lists of varying lengths using a single context. Each list was randomly generated holding m constant for each pattern. Each item was shown to the network once for 6 cycles each; that is, presentation was perfectly sequential. After all the items were learned, the gain was set to 40 and the network's memory was tested by presenting each input pattern (content plus context) and examining the network's output. Table 11.2 presents, for each list length, the total best match for all the outputs and the total number of output units that were spuriously activated across the entire test list (activations greater than .2). Also shown are the expected number of spuriously activated bits as calculated by multiplying the expected number of spurious bits for one pattern ($[n - m]p_{on}{}^m$; see the Appendix) by the list length. For list lengths 20 through 80, the observed number of spurious activations was close to the expected number. For list length 100 however, many fewer were observed (138) than were expected (232). These data support our contention that a Willshaw network can retain a large number of items even when they are presented sequentially in a single context.

TABLE 11.2
Total Best Match and Total Number of Spurious Bits Observed and Predicted as a
Function of List Length

List Length	Total Best Match	Total Spurious Activation	
		Observed	Expected
20	20	0	0.19
40	40	0	4.8
60	58	29	29
80	78	71	96
100	99	138	232

Discrimination

One problem associated with auto-association is that networks can succeed at the task by learning a simple identity transformation without extracting any structure from the pattern set. They can learn to copy the input onto the output. Because our learning scheme mimics that of Willshaw's network, our network always extracts relations among features. For this reason our network is excellent at discriminating between old and new items. Even after learning long lists of items, the system rarely turned on an output unit when presented with a new input. This is true whether the new input consisted of a new context, a new content, or both. Of course, if any of the output units happened to have previously been associated with all the new active content units in contexts that overlapped with all features of the new context, that output unit was spuriously turned on.

Ratcliff (1989) has argued that, unlike human subjects, parallel, distributed models of memory fail to show a positive relation between old/new discrimination and the number of learning trials on old items under conditions of sequential learning. As long as discrimination has not reached asymptote, our system does show such a relation. After pretraining a network with 100 items in two contexts each, we taught it 10 of those items (randomly chosen) in a new context, using a constant gain of 2, and examined the network's ability to reconstruct old items as a function of the number of learning cycles on the items (1, 3, or 5). The experiment was replicated 11 times. When items were shown to the network only *once* and then tested, the mean number of best matches with respect to other items within the set of 10 was 3.3. After each item had been presented three times (item 1, item 1, item 1, item 2, item 2, item 2, ...), mean performance was 7.9. After 5 sequential presentations, all 10 items in every replication were perfectly learned by the best match criterion. This increase in reconstructive ability was highly significant, $F(2,30) = 79.7$; $p < .0001$. Not any of 10 different pretrained items presented in the new contexts scored a best match. In summary, the more learning trials on an item, the better it was reproduced, with little effect on the likelihood of reproducing new items.

Relation between Context and Capacity

Because there are e orthogonal vectors in e episodic units, the context units increase the capacity of Willshaw's network by a factor of e. Items learned in orthogonal contexts cannot interfere with each other. There is no overlap amongst the features of orthogonal contexts, thus there can be no accidental spread of activation to hidden units. We taught a network 50 patterns in a single context, followed by a different 50 in an orthogonal context. After learning the second 50, the network turned on precisely the same output units when tested with the first 50 as it had before learning the second set. Similarly, the first 50 had no effect on learning of the second 50.

Paired-Associate Interference

McCloskey and Cohen found catastrophic interference in a simulation of an experiment by Barnes and Underwood (1959). We show that our network does not suffer from catastrophic interference in this situation.

Using an A-B, A-C design, Barnes and Underwood showed their subjects lists of eight word pairs. The stimulus terms of the pairs were three letter nonsense syllables and the responses were two syllable adjectives. Learning used the method of anticipation: On each trial, A was presented and subjects tried to recall the corresponding response word, which they were then shown. A-B learning progressed in this fashion until criterion performance of one perfect recitation had been reached. A-C learning then took place for either 1, 5, 10, or 20 learning trials. A final cued recall test followed in which subjects were shown each A word and asked to write down both the corresponding B (List 1) and C (List 2) word and to identify the list on which each word appeared.

The results of the Barnes and Underwood study appear in Fig. 11.6 (taken from Fig. 1 of the original paper). The number of B and C words recalled is plotted as a function of the number of trials on the A-C list. The dashed curve shows, not surprisingly, that recall of C items increased over A-C learning trials. The solid curve shows a gradual decline in A-B performance to 52% after 20 trials of A-C learning. Although there was some retroactive interference of A-C learning on recall of B (unlearning), about half the B items were still recalled even after a large amount of A-C learning.

To model paired-associate learning, we assume that individual words are known prior to the experiment, and that nonsense words are interpreted in terms of previously known feature to feature relations. Paired-associate learning consists, we assume, of associating previously learned relations in an experimental context. We therefore pretrained our network on a set of single ($m = 4$) items

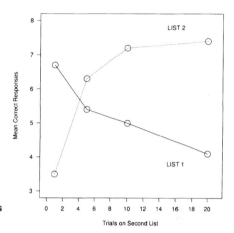

FIG. 11.6.   Data   from   Barnes and Underwood (1959).

before choosing some of those items to form paired-associates. To form a pair, two pretrained patterns (stimuli and responses) whose features did not overlap were chosen and superimposed to create a string having $2m$ active bits. The network was then taught the superimposed strings by auto-association in some episodic context. To simulate the cued recall test, we presented the network with the m bits of each stimulus in the same context. The network was always able to reconstruct the entire pair at output. Except for some spurious output activations due to interference from other patterns, pattern completion performance was perfect. Using superimposition as a model of paired-associates, the system can be pretrained using only individual items; pairs of items need never be shown. Another advantage of this technique is that it generalizes in an obvious way from pairs of items to triplets and to any number of elements, limited only by the choice of n and m. Notice that no distinction is made between backward and forward associations (an assumption that may be justified only for experiments using a small number of learning trials).

To simulate preexperimental learning in the Barnes and Underwood study, we pretrained a network on 100 randomly generated items in two contexts each (using 40 different contexts). Items were shown sequentially for 10 cycles. To model the experiment itself, we chose 24 of these 100 items at random and divided them into three groups of eight items each (A, B, and C). Each B and C item was paired with an A item, under the constraint that no active bits were shared between members of a pair, to form eight A-B and eight A-C pairs. Overlap between active bits of corresponding B and C items was rare, and no more than the overlap between items from different triplets. List 1 learning took place using a single, new context. The eight A-B pairs were presented, one through eight, for several cycles until they all reproduced themselves as measured by the best match criterion. The gain was set relatively low for the first

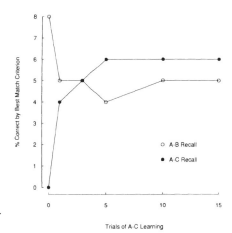

FIG. 11.7. Barnes and Underwood (1959) simulation.

cycle through the set (1.04) and was gradually increased through subsequent cycles. A new context was chosen for the second list that overlapped in one out of three active bits with that of the first list. Note that, holding the similarity between B and C responses constant, the amount of interference between the two lists will be proportional to the similarity between the two contexts. The set of A-C items was cycled through for varying numbers of trials using the same gain schedule.

Cued recall was tested by presenting each A item in the appropriate context and comparing the output against all 16 A-B and A-C pairs. Figure 11.7 shows performance as measured by best match on both A-B and A-C recall as a function of the number of A-C learning trials. Not unexpectedly, A-C learning increased in proportion to the number of A-C trials, reaching asymptote at 6 of 8. Like Barnes and Underwood's subjects, A-B performance did decrease over the first few trials, but not catastrophically, reaching asymptote at 5.

CONCLUSION

Without constraints on stimuli or architectures, catastrophic interference can be a problem for distributed models of memory. Considerations of the structure of natural stimuli, both the structure inherent in natural categories and the independence of arbitrary experiences to one another, reduces the problem appreciably. But the ability of the human memory system to remember and discriminate among arbitrary episodes that are experienced once, briefly, and that are often highly related to other experiences motivates us to suggest architectural constraints that further reduce the problem. We use a network originally proposed by

Willshaw (1981) and introduce the notion of episodic gating, a constraint that goes a long way toward solving the problem and that is consistent with a sizable body of psychological data.

ACKNOWLEDGMENTS

Gordon Bower, Ben Murdock, Charles Rosenberg, and the PDP lab group at Stanford were an invaluble source of comments, help, and ideas.

REFERENCES

Anderson, J. A., Silverstein, J. W., Ritz, S. A., & Jones, R. S. (1977). Distinctive features, categorical perception, and probability learning: Some applications of a neural model. *Psychological Review, 84,* 413–451.

Anderson, J. R., & Bower, G. H. (1972). Recognition and retrieval processes in free recall. *Psychological Review, 79,* 97–123.

Baldi, P., & Hornik, K. (1989). Neural networks and principal component analysis: Learning from examples without local minima. *Neural Networks, 2,* 53–58.

Barnes, J. M., & Underwood, B. J. (1959). "Fate" of first-list associations in transfer theory. *Journal of Experimental Psychology, 58,* 97–105.

Bartlett, F. C. (1932). *Remembering.* Cambridge: Cambridge University Press.

Bransford, J. D., & Johnson, M. K. (1972). Contextual prerequisites for understanding: Some investigations of comprehension and recall. *Journal of Verbal Learning and Verbal Behavior, 11,* 717–726.

Brousse, O., & Smolensky, P. (1989). Virtual memories and massive generalization in connectionist combinatorial learning. *The 11th Annual Conference of the Cognitive Science Society,* 380–387.

Cermak, L. S. (1984). The episodic-semantic distinction in amnesia. In L. R. Squire & N. Butters (Eds.), *Neuropsychology of memory.* New York: Guilford Press.

Craik, F. I. M., & Lockhart, R. S. (1972). Levels of processing: A framework for memory research. *Journal of Verbal Learning and Verbal Behavior, 11,* 671–684.

Deffenbacher, K. A., Carr, T. H., & Leu, J. R. (1981). Memory for words, pictures, and faces: Retroactive interference, forgetting, and reminiscence. *Journal of Experimental Psychology: Human Learning and Memory, 7,* 299–305.

Eich, J. M. (1982). A composite holographic associative recall model. *Psychological Review, 89,* 627–661.

Garskof, B. E., & Sandak, J. M. (1964). Unlearning in recognition memory. *Psychonomic Science, 1,* 197–198.

Graf, P., & Schacter, D. L. (1987). Selective effects of proactive and retroactive interference on implicit and explicit memory for new associations. *Journal of Experimental Psychology: Learning, Memory, and Cognition, 13,* 45–53.

Graf, P., Squire, L. R., & Mandler, G. (1984). The information that amnesic patients do not forget. *Journal of Experimental Psychology: Learning, Memory, and Cognition, 10,* 164–178.

Hetherington, P. A., & Seidenberg, M. S. (1989). Is there "catastrophic interference" in connectionist networks? *Proceedings of the Cognitive Science Society,* 26–33.

Hinton, G. E., & Plaut, D. C. (1987). Using fast weights to deblur old memories. *Proceedings of the Cognitive Science Society,* 177–186.

Humphreys, M. S., Bain, J. D., & Pike, R. (1989). Different ways to cue a coherent memory system: A theory for episodic, semantic, and procedural tasks. *Psychological Review, 96,* 208–233.

Hyde, T. S., & Jenkins, J. J. (1969). Differential effects of incidental tasks on the organization of recall of a list of highly associated words. *Journal of Experimental Psychology, 82,* 472–481.

Jacoby, L. L., & Witherspoon, D. (1982). Remembering without awareness. *Canadian Journal of Psychology, 36,* 300–324.

Kohonen, T. (1988). *Self-Organization and Associative Memory, Second Edition.* Berlin: Springer-Verlag.

Lewandowsky, S., & Murdock, B. B. (in press). A distributed memory model for associative learning. In D. Vickers & P. L. Smith (Eds.), *Human information processing: Measures, mechanisms, and models.* Amsterdam: North-Holland.

McClelland, J. L. (1986). Resource requirements of standard and programmable nets. In D. E. Rumelhart & J. L. McClelland (Eds.), *Parallel distributed processing, Vol. 1* (460–487). Cambridge, MA: MIT Press.

McClelland, J. L., & Rumelhart, D. E. (1985). Distributed memory and the representation of general and specific information. *Journal of Experimental Psychology: General, 114,* 159–188.

McCloskey, M., & Cohen, N. J. (in press). Catastrophic interference in connectionist networks: The sequential learning problem. In G. B. Bower (Ed.), *The psychology of learning and motivation.*

Mensink, G. J., & Raaijmakers, J. G. W. (1988). A model for interference and forgetting. *Psychological Review, 95,* 434–455.

Mishkin, M., Malamut, B., & Bachevalier, J. (1984). Memories and habits: Two neural systems. In G. Lynch, J. L. McGaugh, & N. M. Weinberger (Eds.), *Neurobiology of learning and memory* (pp. 65–77). New York: Guilford Press.

Moscovitch, M. (1982). Multiple dissociations of function in amnesia. In L. S. Cermak (Ed.), *Human Memory and Amnesia* (337–370). Hillsdale, NJ: Lawrence Erlbaum Associates.

Murdock, B. B. (1982). A theory for the storage and retrieval of item and associative information. *Psychological Review, 89,* 609–626.

Ostergaard, A. L. (1987). Episodic, semantic and procedural memory in a case of amnesia at an early age. *Neuropsychologia, 25,* 341–357.

Parkin, A. (1986). Residual learning capability in organic amnesia. *Cortex, 18,* 417–440.

Postman, L., & Stark, K. (1969). Role of response availability in transfer and interference. *Journal of Experimental Psychology, 79,* 168–177.

Ratcliff, R. (1989). *Connectionist models of recognition memory: Constraints imposed by learning and forgetting functions.* Manuscript submitted for publication.

Rumelhart, D. E., Hinton, G. E., & Williams, R. J. (1986). Learning integral representations by error propagation. In D. E. Rumelhart, & J. L. McClelland (Eds.), *Parallel distributed processing, Vol. 1* (318–362). Cambridge, MA: MIT Press.

Rumelhart, D. E., & McClelland, J. L. (1986). *Parallel distributed processing, Vol. 1: Explorations in the microstructure of cognition.* Cambridge, MA: MIT Press.

Schacter, D. L. (1985). Priming of old and new knowledge in amnesic patients and normal subjects. *Annals of the New York Academy of Sciences, 444,* 41–53.

Schacter, D. L., & Tulving, E. (1982). Memory, amnesia, and the episodic/semantic distinction. In R. L. Isaacson, & N. E. Spear (Eds.), *The expression of knowledge* (pp. 33–65). New York: Plenum Press.

Sloman, S. A., Hayman, C. A. G., Ohta, N., Law, J., & Tulving, E. (1988). Forgetting in fragment completion. *Journal of Experimental Psychology: Learning, Memory, and Cognition, 14,* 223–239.

Squire, L. R., & Cohen, N. J. (1984). Human memory and amnesia. In G. Lynch, J. L. McGaugh, & N. M. Weinberger (Eds.), *Neurobiology of Learning and Memory* (pp. 3–64). New York: Guilford Press.

Sutton, R. S. (1986). Steepest-descent learning procedures for networks. *Proceedings of the Cognitive Science Society, 823–831.*

Tulving, E. (1972). Episodic and semantic memory. In E. Tulving & W. Donaldson (Eds.), *Organization of memory* (pp. 381–403). New York: Academic Press.

Tulving, E., Schacter, D. L., & Stark, H. A. (1982). Priming effects in word-fragment completion are independent of recognition memory. *Journal of Experimental Psychology: Learning, Memory, and Cognition, 8,* 336–342.

Tulving, E., & Thornton, G. B. (1959). Interaction between proaction and retroaction in short-term retention. *Canadian Journal of Psychology, 13,* 255–265.

Warrington, E. K., & Weiskrantz, L. (1974). The effect of prior learning on subsequent retention in amnesic patients. *Neuropsychologia, 12,* 419–428.

Willshaw, D. (1981). Holography, Associative memory, and inductive generalization. In G. E. Hinton & J. A. Anderson (Eds.), *Parallel models of associative memory* (pp. 83–104). Hillsdale, NJ: Lawrence Erlbaum Associates.

APPENDIX

Following McClelland (1986), we derive the capacity of Willshaw's network. We assume that patterns are generated randomly and with equal probability. Recall that m_1 is the number of 1 bits in n_1 input units and m_2 is the number of 1 bits in n_2 output units. The storage of each pattern requires setting $m_1 m_2$ of the $n_1 n_2$ weights in the network to 1. The probability that a weight is not turned on during learning of a single pattern is therefore $1 - m_1 m_2 / n_1 n_2$ and the probability that any given weight connecting an input to an output has been set to 1 after learning r patterns is

$$p_{on} = 1 - (1 - m_1 m_2 / n_1 n_2)^r. \tag{3}$$

An output unit will turn on spuriously if and only if the weights connecting it to all activated input units are 1, which occurs with probability $p_{on}{}^{m_1}$. As this could occur for any of the $n_2 - m_2$ output units not in the output pattern, the average number of spurious output bits upon presentation of an input is

$$(n_2 - m_2) p_{on}{}^{m_1}$$

which is less than 1 if

$$p_{on} \le [1/(n_2 - m_2)]^{1/m_1}. \tag{4}$$

Substituting (3) into (4) and recombining gives

$$(1 - m_1 m_2 / n_1 n_2)^r \ge 1 - [1/(n_2 - m_2)]^{1/m_1}.$$

If $m_1 m_2 / n_1 n_2 < .1$, that is, if patterns are sufficiently sparse, then

$$\log(1 - m_1 m_2 / n_1 n_2)^r \approx -r m_1 m_2 / n_1 n_2$$

so that

$$-r m_1 m_2 / n_1 n_2 \ge \log(1 - [1/(n_2 - m_2)]^{1/m_1})$$

or

$$r \leq -log(1 - [1/(n_2 - m_2)]^{1/m_1})(n_1 n_2 / m_1 m_2).$$

If we choose $m_1 \leq \log_2(n_2 - m_2)$ then .69 is a lowerbound for $-log(1 - [1/(n_2 - m_2)]^{1/m_1})$ so that

$$r \leq .69 n_1 n_2 / m_1 m_2.$$

The identical analysis applies to the case of an auto-associator ($m_1 = m_2 = m$ and $n_1 = n_2 = n$) with the result that

$$r \leq .69(n/m)^2.$$

12 What Good is Connectionist Modeling? A Dialogue

Jay G. Rueckl
Stephen M. Kosslyn
Harvard University

One of the remarkable aspects of W. K. Estes' career is that his work has always been current; indeed, he has more than kept up with the field: He has consistently been a leader. Most recently, Bill has become interested in "connectionist" (also known as "parallel distributed processing" or "neural net") models. We decided to reflect on these models as a tribute to Bill's work, and also as a demonstration of its continuing vitality. When we tried to outline this chapter, we discovered that we disagreed on fundamental points, and were not sure where Bill would fall on the issues. Thus, instead of trying to paper over our differences (which would leave a lumpy wall indeed!), we decided to air them, and perhaps draw Bill into yet another controversy. We felt that not only would we benefit from his wise reflections and broad perspective on these matters, but so would the field as a whole. Thus, we herald not the end of a career, but the beginning of a teaching- and committee-free phase of a career; we acknowledge not only a history of major contributions, but also a continuing role in the community. And we thank Bill not only for his generous spirit and steady hand as a colleague, but also for his guidance and patience. Bill has been a marvelous person to have in one's department, and we look forward to many more years of his company.

KOSSLYN: Let me try to get to the heart of the major issue that divides us, and perhaps much of the field. It seems to me that connectionist networks are not to be taken seriously as process models of higher level cognitive tasks, such as reading. Rather, there are only two reasons to implement such connectionist models: First, to discover how difficult it is to take an input and map it to a specific output in a certain class of brain-relevant (more on this shortly) architectures, and, second, to discover what information in the input allows it to be mapped to the output.

RUECKL: Most connectionists, myself included, believe that at least some connectionist networks can be taken seriously as process models. The idea is that many aspects of network behavior—learning, interference and transfer effects, automatic generalization, differences in performance as a function of stimulus frequency or typicality—map rather directly onto human behavior. So your claim, that the only reasons to simulate a connectionist model are to learn about how difficult a mapping is and what information is used in that mapping, is rather surprising. I'm curious about why you think simulations tell us about those aspects of an input/output mapping, and conversely, why you think they don't tell us about anything else. Why don't we start with your first point, that simulations provide a measure of the difficulty of an input/output mapping.

KOSSLYN: For present discussion, let's focus on simple feedforward, back propagation models (see Rumelhart, Hinton, & Williams, 1986). These are not only very simple, but also probably the most popular variety (maybe in part because they are so simple). Such a network simulation is a kind of "poor man's way" of assessing computational complexity. One can discover how difficult an input/output mapping is by determining the number of training trials required to establish the mapping or the amount of error in the mapping after a fixed number of training trials. An example of this use was presented by Rueckl, Cave and Kosslyn (1989), who found that a split network could map both the identity and location of a shape more efficiently (i.e., with less training) than an unsplit net—which was intriguing because the brain has separate pathways of achieving each aspect of this mapping (e.g., Ungerleider & Mishkin, 1982).

RUECKL: What is the significance of knowing how difficult a particular mapping function is?

KOSSLYN: Because one can ask the following kind of question: Why did the brain separate processing of location from shape? Or, why didn't it separate processing of location and size? One can ask questions about the segregation and organization of information processing. One then can investigate the question using two paradigms. The first was that adopted by Rueckl et al. In that paradigm, a single network that maps various kinds of input to output is compared to a split network, with different partitions performing different aspects of the mapping. If the split net is more efficient computationally, that is an argument as to why the brain evolved separate subsystems. By examining the network, one may be able to argue, for example, that location and shape are qualitatively different kinds of functions, whereas location and size (for reasons that may not be intuitively obvious) share some aspects of the mapping, and so it is more perspicuous to have those mappings done by a single network.

A second paradigm is to train a network on one mapping and then transfer it to another. If one finds positive transfer, this result suggests that the mappings share computations. Alternatively, if one finds interference, this suggests that the mappings are distinct. We (Kosslyn, Chabris, Marsolek, & Koenig, in press) have performed such experiments in order to examine the distinction between "categorical" (e.g., left/right, above/below) and "coordinate" (e.g., 1.3 inches) spatial relations (Kosslyn, 1987), and found evidence for the distinction between the two kinds of computations.

RUECKL: In the first paradigm, the simulations are used to measure the difficulty of a mapping, and conclusions are drawn from there. But in the second paradigm, transfer effects seem to say less about mapping difficulty than about the kinds of information each mapping uses—the second point you started with. Before pursuing that line of thought, are there any other reasons to determine how difficult a particular mapping function is?

KOSSLYN: Another reason is that certain kinds of mappings may be wildly implausible. If it turns out that it is virtually impossible to map an input to an output within this kind of architecture, that means that the problem must be broken into subproblems and/or restated from a different perspective.

RUECKL: That reminds me a bit of Minsky and Papert's (1969) work. They provided analytic proofs that certain classes of networks couldn't perform certain mappings in plausible ways. As a result, connectionism almost ended in an early grave.

Let me suggest another reason why determining the complexity of a mapping might be useful. Suppose one finds differences in network behavior as a function of the complexity of the mappings the networks are computing. One might ask if biological systems that are thought to compute those functions differ behaviorally in the same way. The idea is that psychological predictions can be generated on the basis of complexity. For example, if the complexity of a function is measured by the number of learning trials needed for performance to reach some criterion, then one factor that influences complexity is something we've called "systematicity" (Rueckl et al., 1989; Rueckl, Kolodny, & McPeek, 1990). Systematicity refers to the degree to which similar input patterns are mapped onto similar output patterns. The more systematic a mapping—that is, the more likely it is that similar input patterns are mapped onto similar output patterns—the easier it is for a network to learn. For example, in the what/where simulations mentioned earlier, the where problem was much more systematic than the what problem, and it was also much easier to learn.

It turns out that the behavior of a network varies with the systematicity of the mapping function being computed. In a large set of simulations, we've shown that networks that compute systematic mappings not only learn more quickly, but are also less susceptible to retroactive and proactive interference and better at generalizing to unfamiliar input patterns (Rueckl et al., 1990). If, by hypothesis, a biological system computes two functions that differ in systematicity, then one would expect to find behavioral differences along the same lines. That is, one would expect the process computing the more systematic mapping to be less susceptible to interference and better able to respond to a novel input.

KOSSLYN: In principle that sounds fine. The trouble is that we do not know the proper way to represent the input or output. And in my view if one doesn't have these representations down, one can't say anything—because all one is studying are input/output mappings. So systematicity is going to hinge on the formalisms used to define the input and output vectors. Thus, unless one knows the way the brain is coding the stimuli, one really can't say much of anything about systematicity, as far as I can see.

RUECKL: But unless you know the proper way to represent the input or output, what can the simulations tell you about computational complexity? The difficulty of learning a mapping is just as dependent on the representational formalisms as any other aspect of a network's behavior.

KOSSLYN: I have two responses to this concern: First, the input and output are kept constant in the kind of paradigms I advocate. In the split/unsplit paradigm, one examines differences in the internal architecture. Thus, the results do not hinge on relations among different input representations and among different output representations, as is the case in your studies of systematicity; all of that is held constant in both paradigms I advocate.

This leads into my second point, that networks can tell one about which aspects of the input are used to establish the mapping. One way to check that the input/output representations capture critical aspects of the input to a real neural network is to analyze the "receptive" and "projective" fields of the hidden units. The receptive field can be determined by discovering which input units activate a given hidden unit maximally, whereas the projective fields are determined by examining the weights on the connections from a hidden unit to each output unit. The hidden units perform a kind of nonlinear multiple regression analysis, with the weights on their connections reflecting the relative importance of different aspects of the input for the mapping. So, for example, in Rueckl et al. (1989) we found that the hidden units used to map "what" information in a split network had some receptive fields that corresponded to vertical stripes, horizontal stripes, and so on, whereas the hidden units used to map "where" information had receptive fields that corresponded to the left, right, top, or bottom of the input array. Different combinations of these hidden units would be activated by a given input, allowing the proper output to be produced. By analyzing the patterns of weights on the connections, we discovered how the mapping was accomplished, which converged with facts we know about how the brain accomplishes the analogous mappings. Thus, the representations (patterns of weights on connections) formed to map input to output gain plausibility as idealizations (not detailed models) of the corresponding biological representations.

RUECKL: At some level our approaches seem quite similar. We both advocate carrying out simulations where some aspects of the model are held constant and other aspects are varied. And thus, the value of the simulations depends on having made good choices about the factors that were held constant. Moreover, we both rely on converging evidence to justify those choices. In the approach you're advocating the converging evidence comes from examining the receptive and projective properties of the hidden nodes and finding that they make use of the same kinds of information that are known to be used by the brain in carrying out an analogous mapping. In the approach I'm advocating, the converging evidence comes from comparing the behavior of the system to human behavior. From the basic assumptions of the model a set of predictions can be derived about the situations under which one should expect to find positive or negative interference, generalization, and so forth. Finding empirical support for these predictions suggests that the basic assumptions are on the right track. Conversely, the empirical failure of these predictions suggests that the brain doesn't perform the computations as claimed. Either way, one now has a set of empirical constraints to shape the further specification of the theory.

KOSSLYN: That's true, but what you've done is to point out that there are some behaviors that offer constraints for networks. That doesn't say much about what one can get from the network itself.

RUECKL: Well, it's by studying networks that we learn about their behavioral properties in the first place. That's how we found that interference and generalization vary with systematicity.

KOSSLYN: But one does not study networks in general; one studies specific models. How do you know that the specific network you are examining generalizes properly to other ways of representing the input and output? One problem here is that you have to know too much before you can set the network up; if we knew how to represent the input and output properly, we should be able to determine systematicity analytically.

RUECKL: I agree that we need to know how to represent the inputs and outputs properly in order to take the model seriously. But the criteria for what counts as "properly" changes over the course of the development of a model. The way to build a psychological model is to start with a set of fundamental assumptions that can be justified theoretically and empirically. These assumptions can be rather general, and thus together might define a broad class of models. For example, an assumption common to an entire class of models might be that the representation of the inputs is such that it captures some similarity metric. There are likely to be a number of alternative representational schemes that capture this similarity metric, resulting in a set of competing submodels. To choose between them, one might set up networks that implement each of the alternatives and look for ways that they differ behaviorally. These differences are in effect empirical predictions that can then be tested.

KOSSLYN: If I have understood correctly, you are saying that if one understands the similarity metric defined over the input and output representations, then one is in a position to say something about the input/output mapping. I wonder how one can understand input similarity without observing input/output mappings. Similar inputs produce similar outputs, and dissimilar inputs produce dissimilar outputs. So, observations about the mapping are used to help define similarity, which means that similarity cannot be used to provide an independent account of that mapping.

RUECKL: The claim isn't that similarity provides an account of the mapping—similarity is part of the definition of the mapping. Instead, the claim is that we can use simulations of alternative theories of the mapping to generate empirical predictions. It's the empirical data that provide evidence for or against a particular theory.

KOSSLYN: The trouble is, there are too many degrees of freedom. The way the input is represented is only one of the things one can manipulate in these models. One can vary the actual mapping apparatus in terms of the number of hidden units, the pattern of connectivity, the structure in the connections one specifies ahead of time, and, of course, the input and output representations. And then there are all of the other parameters concerned with "learning" rate. There are a very large number of ways of varying a network, once one starts thinking about combinations of different properties (cf. Massaro, 1988). So I don't think much can be learned simply by looking at variations among the input and describing the consequent mappings. The ease of mapping in a given network depends on all these factors. In the work we do, we hold all of these things constant when comparing split and unsplit nets, or when examining transfer.

RUECKL: The assumption is that theoretically uninteresting factors don't interact in complex and unknown ways. On the other hand, interactions among the theoretically interesting factors (e.g., the representation of the inputs and outputs, the patterns of connectivity, and so on) are just what we're looking for. They serve as the basis for empirical predictions and allow for further refinement of the model.

KOSSLYN: That's quite a big assumption. I don't think one can purchase that on faith; one needs to buttress it with exactly the sort of massive parametric variations that nobody wants to do. People almost never orthogonally vary characteristics of the models when testing them. That is, if they vary the hidden units, they'll also vary the representation, or the connectivity, or something. So one doesn't really know what to make of results from connectionist modeling because researchers typically don't really shake it out, telling one what the critical characteristics are in a given domain. In particular, researchers rarely explore the importance of different input representations, which I think is absolutely critical. As far as I know, no one has been systematically exploring the importance of different kinds of input representations for specific mappings.

RUECKL: Actually, I think we're now starting to see studies involving the systematic manipulation of certain variables. Often these variables are parameters that define the structure or processing characteristics of a network (see, for example, Ahmad & Tesauro, 1989; Baum & Haussler, 1989; Kolen & Pollack, 1990; Plaut, Nowlan, & Hinton, 1986). However, in a number of studies the effects of differences in the representational format and other aspects of a network's training environment have been explored (e.g., Ahmad & Tesauro, 1989; Hetherington & Seidenberg, 1989; Rueckl et al., 1990).

Still, I'd agree that the consequences of choosing a particular representational scheme is an issue that deserves more attention. In fact, I think that to some extent it's forcing itself upon us. For example, there is an ongoing debate about whether connectionist networks are too susceptible to interference to model episodic memory in humans. One of the key insights emerging from this debate is that the amount of interference exhibited by a network depends crucially on how the inputs and outputs are represented (cf. Brousse & Smolensky, 1989; Kortge, 1990; McCloskey & Cohen, 1989; Ratcliff, 1990). Thus, any conclusions one might draw about the viability of connectionist models of human memory rest on underlying assumptions about how the inputs and outputs of those models are represented.

I'd like to return to one aspect of your argument that I find puzzling. It seems that the same factors that could influence, say, interference or automatic generalization, could also influence the relative difficulty of various input/output mappings. Why is this a problem in one case but not the other? For example, suppose computational complexity varies as a function of the number of hidden units. What then?

KOSSLYN: Clearly, there will be a saturation point at which there are too few hidden units, and the mapping cannot be achieved; and there may be a point at which there are so many hidden units that the net can map many functions equally easily (one can think of this as a ceiling effect). That's fine, but in the range between those two extremes, if the results vary significantly depending on the number of hidden units and so on, then one is in trouble; one cannot make claims about the relative difficulty of a mapping in different kinds of nets. However, one must have theoretical reasons for manipulating properties of networks such as connectivity patterns. For example, one splits the internal connections in a network because one is examining a theory.

It is true that we also must vary a number of factors (particularly the number of hidden units) to determine whether a result has some generality, but we need not cover the entire search space. And we don't care if the various factors interact in complex ways, so long as the relative difference in the ease of establishing the input/output mapping is preserved. But your use of networks hinges critically on the assumption that they do not interact.

RUECKL: Not necessarily. If the behavior of a network is robust over a range of values along some dimension, that's fine. It means we can generalize from our results in a straightforward way. But if the behavior of a network changes as a function of some variable, that's fine too, because it allows for further refinement of the model. That is what I was trying to get at earlier with the example about varying representational formalisms. If the implementation of theoretically interesting variables (like how similarity metrics are embodied in the representations of the input and output vectors) influence the behavior of the network, then by manipulating them the space of possible configurations can be partitioned into a number of equivalence regions, each of which is characterized by an ensemble of behavioral properties. We can then try to determine which regions are consistent with human or animal behavior and which regions are not.

KOSSLYN: I am a little confused; you are agreeing that one must consider every possible combination of variables to draw theoretical inferences? This sounds as if you are simply assuming that things will work out okay eventually, and are offering promissory notes for now.

RUECKL: Well, it's always the case that in some sense all possible combinations of variables are relevant, and that's true for any sort of theoretical endeavor, connectionist or not. But it's also the case that for practical matters we would like to narrow down the search space if possible. What I'm claiming is that we can narrow it down. Some variables are theoretically uninteresting, and if previous work indicates that the behavior of a network is stable over a range of values on such a variable, then we are justified in fixing that variable at a reasonable value. Other variables may be theoretically interesting, but only in certain contexts. That is, a variable may not matter with regard to a certain aspect of behavior, but it may be crucial for determining other aspects of behavior.

Perhaps what makes it seem like a bunch of promissory notes is that connectionism is a young field. We're still learning what networks can or can't do, and about how a variety of factors influence their behavior.

KOSSLYN: What you say seems reasonable, but I still have a worry: One cannot easily generalize from results showing stability along a given dimension when those networks had different values along other variables; perhaps interactions only occur in restricted ranges of variable values. If every possible variable may interact, then it is going to be very difficult to accumulate theory. One is building a boat at sea. This reminds me of Marr's (1982) arguments for the necessity of a "theory of a computation." Without sufficiently strong prior constraints, motivated decisions about the model, the results from experimenting with networks may not mean very much. By analogy, if one had a model airplane in a wind tunnel, but one did not have a theory that told one to vary the shape of the wings, I wouldn't particularly want to get an airplane that was designed after one fooled around with the model and found the configuration that seemed to fly best. I want a theory about why that range of curvature produces the following effects rather than another, and so on, otherwise I won't have any confidence that the behavior of the model will generalize properly to that of the airplane.

RUECKL: I agree that models need to be constrained and motivated in principled ways. What I'm suggesting is that there are both top-down and bottom-up aspects of theory construction. In addition, the notion of levels of specificity is important. In the process of constructing a theory we make increasingly specific hypotheses about the processes under consideration. We start out, from the top down, with a set of well-motivated fundamental hypotheses. These hypotheses don't necessarily include everything that one would expect to end up with in a theory, but they include some crucial assumptions that can be built upon.

KOSSLYN: And that is the nut of our disagreement. Of course we both agree one has to have *some* top-down basis for formulating a model. But I think that we just don't know enough, a priori, to structure networks that can be taken seriously as process models. The equivalence class of these models is so huge that we don't know exactly what defines the critical properties that make a model a member of this class and the properties that are irrelevant. That is a major reason why I think we have to stay at a more abstract level. I don't think we know enough to be able to specify that it's the shape of the wings that's critical.

RUECKL: I think that's one of the reasons we do simulations. We're trying to come to an understanding of how and why networks behave the way they do. We're also trying to understand how the behavior of those networks maps onto human behavior.

KOSSLYN: But I don't think one can do that bottom-up. Let's stick with the airplane example for another minute. I wonder whether one could ever inductively figure out the best way to build an actual airplane simply by haphazardly playing around with shapes of model airplanes. I think one has to have a theory, an analytic theory, that tells one what's going to be important. Nelson Goodman (1973) wrote a book about induction called *Fact, Fiction and Forecast*. One of his major points is that without a lot of prior structure, one

can't ever generalize from examples because there are just too many characteristics that vary at once. And one doesn't know which of them are going to
be relevant for the necessary comparison when one looks at successive examples. One has to have prior structure. My view is that the current state of the
network world is such that *no* amount of messing around bottom-up, with a
strategy of let's-vary-this-vary-that-try-it-and-see-what-happens, is going to
allow us to inductively generalize, so that we can take a network seriously as a
process model.

RUECKL: I'm not sure that the situation is as bad as you're making it out to
be. It's just not the case that we have no understanding of what's relevant or
irrelevant to network behavior. For example, we know that networks tend to
learn faster but generalize worse with more hidden units (Baum & Haussler,
1989; Plaut et al., 1986); we know that increasing the learning parameter
reduces proactive interference, but increases retroactive interference
(Schneider & Detweiler, 1988); we know that learning rate and interference
vary with systematicity (Rueckl et al., 1990); we know that there is a class of
problems that can't be computed by single-layer feed forward networks (Minsky & Papert, 1969); we know that given certain constraints a recurrent
network will perform a gradient descent search (Hopfield, 1982; Hinton &
Sejnowski, 1986). These are the sort of basic principles that can allow us to
generalize from particular simulations. Granted, we don't know everything
there is to know about network behavior. But it seems to me that it's by having
a community of people carrying out simulations that we generate the data
needed to guide the search for these principles. The simulations can tell us
what might be important.

KOSSLYN: I agree that progress is being made, but it seems to be at exactly
the level I advocate: considering networks as input/output mapping devices.
The results you cited do not offer any particular support for taking networks
seriously as process models: The operation of any specific model depends
critically on the choice of particular input and output representations.

RUECKL: Well, part of your argument against taking connectionist networks as process models is that we don't understand the behavior of those
networks well enough to generalize from particular simulations. What I was
arguing is that we do not understand their behavior well enough to justify
certain generalizations. If one accepts that assertion, the case against taking
networks as process models is weakened.

  I do appreciate your point, however. In many connectionist applications
the choice of the representational format often seems to be completely arbitrary, and one has to wonder how the behavior of the system would change if a
different representational scheme was used. And there is no denying that the
behavior of a connectionist network depends on how its inputs and outputs are
represented. But I don't take this as implying that connectionist networks

should not be taken as process models. Instead, the approach I'm advocating takes advantage of the importance of the representational scheme. The idea is to examine the behavior of systems that employ different representational schemes, and to then use relevant data about human behavior in order to evaluate these schemes.

One point this brings out is that in this approach there is a close tie between simulation results and the results of empirical studies. Earlier, you mentioned that one of the things we can learn from simulations is what information is used in carrying out a mapping, and you seemed to be linking simulation results to the results of neurophysiological studies. Could you say more about how simulations can tell us what information is used to carry out a mapping?

KOSSLYN: The idea is that the network performs something like a regression analysis, where it figures out what the independent variables are. This kind of analysis turns out to be interesting because, at least with three-level, feed forward backprop models, the network captures aspects of how neurons perform the mapping. Two examples convinced me: Andersen and Zipser (1988) found that the receptive fields developed by the hidden units of their network captured many of the properties of actual receptive fields in the part of the brain they were modeling, and Lehky and Sejnowski (1988) found that a network developed end-stopped units after it was trained to recover shape from shading (this last result is particularly impressive because it was unclear what such neurons could be doing). Frankly, I find these results utterly astonishing, but it seems to be the case that the networks capture something about the kinds of mappings carried out by the brain. So it's of interest to look at how those mappings are performed. I think one gets some insight into the nature of how an input/output mapping is performed, which then can inspire one to think about process models. But the network is still one level more abstract than a process model per se.

RUECKL: Earlier you said you don't think one should take hidden units seriously. I'm trying to reconcile that with what you just said.

KOSSLYN: In general, I am reluctant to take the hidden units seriously as corresponding to individual representations in the brain; I want to think of the things on a more abstract level, as performing a mapping from input to output. The hidden units tell one something about how that mapping is performed, but that doesn't imply that there must be representations like those formed by the hidden units in the brain somewhere. These representations might be implemented in numerous ways; for example, the corresponding representation of the brain may be a population of neurons that work together. Or neurons may participate in several mappings, and so have different properties than those found in a network dedicated to performing one mapping. On the other hand, in some cases the network may be set up so that the mapping is in fact similar to that performed by a small group of neurons, and hidden units might reflect some properties of hidden units. This is logically possible, but depends critically on selecting the appropriate mapping.

RUECKL: In some sense the Andersen and Zipser results were compelling because there seemed to be a one-to-one mapping from units to neurons.

KOSSLYN: I agree; either there is something very deep or something very superficial here. These models are members of very large equivalence classes, and I suspect that some simple property—such as gradient descent or the kind of nonlinear threshold accumulation functions used—is what is capturing the brain-like mappings. My lab has done some work that follows up Andersen and Zipser, and we (O'Reilly, Kosslyn, Marsolek, & Chabris, 1990) were able roughly to reconstruct with the hidden units all three types of neurons that Andersen discovered. However, such a 1:1 correspondence between properties of hidden units in a network and actual neurons should occur only if the input and outputs of the network capture critical aspects of the corresponding neural representations.

RUECKL: So even if a one-to-one mapping isn't assumed, the hidden units are still of interest because they will tell us something about what's being abstracted?

KOSSLYN: I should think so, but it's a question of the level of abstraction. One of the nice things about staying relatively close to the brain is that one has a better shot at making sure that one is modeling a distinct neural network; one may actually be modeling a real neural network that's been anatomically identified. In addition, the nature of the input/output to a network can be reasonably clear in the visual system, anyway. One knows what kinds of information feed into some of these areas both from the neuroanatomy (where one discovers what's connected to what) and neurophysiology (where one discovers response properties of neurons). As one studies higher level processes, say, for example, reading, it is much more of a gamble as to what the actual subsystems ought to be and what the representations ought to be like.

RUECKL: I like the approach you're describing, which tries to tightly couple software and wetware. But I'm not sure that it's as well-suited for some cognitive processes as for others. If we had to wait for neuroscientists to tell us what parts of the brain are involved in semantic representation, and worse, how the properties of individual neurons in those areas relate to semantic representation, I suspect that we'd be in for a long, long wait.

It seems to me that for more abstract problems the more appropriate strategy might be to start with very general principles and work downward toward greater and greater levels of specificity. Take visual word identification. The approach would be to first consider the most fundamental properties of connectionist networks—for example, slow incremental learning or distributed representation. One would want to show that what we know about word identification is readily interpreted in terms of those concepts, and one would also want to derive testable empirical predictions from them. For example, in a connectionist system, learning one association influences the computation of other associations, with the transfer being relatively beneficial if the other association is similar to the one just learned and relatively harmful otherwise.

It turns out that in human behavior transfer effects of just this sort occur. For example, it's been found that reading aloud an unfamiliar pseudoword facilitates both the subsequent identification of that pseudoword and the subsequent identification of orthographically similar pseudowords as well (Rueckl, 1990). Before doing the experiments it wasn't obvious that transfer effects of this sort would occur, and in fact most nonconnectionist models predict that they won't. But similarity-dependent transfer does occur, and the way we found that out is by working from a prediction based on the general properties of connection systems. So that's the first step, but it's not the end of the story. The fact that we get similarity priming tells us that similar words are represented in a similar way. The next step is to consider alternative representational schemes that capture similarity in different ways, and to try to find empirical predictions that might tease apart these alternative schemes.

KOSSLYN: This all sounds slightly circular to me, because one defines similarity by the way the vectors are set up for the input. How is the conversation of a word to a binary vector representation determined? It has to be largely arbitrary. By deciding that, one has already defined what's similar and isn't similar. Furthermore, simply by knowing some general properties of the networks one can pretty much guarantee that the network will behave in certain ways by specifying the input representations appropriately. If one makes input representations so that they are similar, one will get interference or facilitation, depending on the paradigm (that is, the kinds of mappings you're trying to achieve). If one made the input representations dissimilar, one would get different results. So, again, this just emphasizes my point that all the theory is really being done at the level of choosing what the network is going to compute (that is, the input/output transformation) and the input and output representations.

RUECKL: But that's not enough. Defining a mapping and the representations of the inputs and outputs of that mapping underdetermine behavior. Will there be interference? Frequency effects? Automatic generalization? It depends on how the mapping is implemented. Different algorithms, connectionist or not, carry out the mapping in different ways, and thus differ along these dimensions.

For example, until recently little research had been done on the similarity-dependent transfer effects I mentioned earlier, probably because the most influential models of word identification did not generate any interest in them. But in the context of the connectionist framework, similarity effects are interesting. Based on the fundamental assumptions of the connectionist framework, we predicted that similarity transfer would occur. The empirical data are consistent with that prediction, and thus provide support for those fundamental assumptions.

KOSSLYN: How many other connectionist experiments produce the right predictions? Tons of them, I would think.

RUECKL: Well, there are tons of nonconnectionist models, too. Sometimes they make the right predictions and sometimes they don't. If the connectionist account of word identification is completely off the mark, we'll find out soon enough, because it will make a series of false predictions. On the other hand, if it consistently makes the right prediction, then it's probably a good model. So goes science.

KOSSLYN: One cannot argue with success, but successful predictions from models don't mean much unless one understands the principles that determine the predictions. For higher level processes, such principles seem to be few and far between, and many people doing that kind of work don't even seem to care about them. They just hack it.

RUECKL: Your point is well taken. Plausible psychological models must be motivated by principled theoretical assumptions. I'd grant that much of the work from the first wave of connectionism didn't satisfy this criterion. One reason was that we were just beginning to cut our teeth, so to speak, and we needed to see if connectionism had any hope at all of working in a wide domain of problem areas. As a result, we ended up constructing toy models in order to demonstrate how connectionism might approach those domains. Much the same thing happened in the early days of classical AI. It's also the case that not all connectionists are psychologists. Many connectionists are interested in building smart machines, period. They aren't concerned with the psychological validity of their systems.

The important thing is that some connectionists are psychologists, and it's up to them to demonstrate that connectionist models are psychologically valid. Stephen Palmer (1987) has made this point quite well. As he argued, in the next stage of connectionism toy models should no longer be acceptable. Instead, connectionists should provide well-motivated, psychologically plausible models and perform the kinds of analyses that provide a solid understanding of why the models behave as they do. An equally important part of their job is to go out and do experiments that test the predictions of the models. If the predictions are correct with reasonable consistency, then we say "Well, it's a good model." If not, we go back to the drawing board. Isn't that true of any model, connectionist or otherwise?

KOSSLYN: When one gets a successful prediction from these models, it is not clear to me that one really knows why one got it. And if one doesn't know why one got it, then one doesn't know what to pass along to the next generation of models, to build on the success. My point earlier was that the behavior of the model depends on many interactions between the input representation, connectivity, and so forth. And unless one knows exactly what it was that resulted in the prediction of interest, it seems to me one doesn't know what one's got.

RUECKL: Well, I'd agree that sometimes what we think is relevant about a connectionist model really isn't, and vice versa. But that's true of nonconnectionist models as well. All in all, I don't think the situation is as bad as you're suggesting. Typically we do have strong candidates for what's theory-relevant and what's theory-irrelevant. Going back to the word identification example, how the inputs and outputs are represented is clearly theory-relevant. That's not the whole story, however. Exactly which aspects of the input and output representations are theory-relevant depends on the set of data you're trying to explain. In the case of similarity-dependent transfer effects, it's likely that any representational scheme that captures orthographic similarity in the inputs and outputs will account for the data. For other phenomena, more specific assertions might be required. For example, accounting for transposition errors (that is, errors where the order of the letters in a word is misperceived—see Estes, 1975) may require more specific assumptions about how the representational scheme captures letter position information.

KOSSLYN: I believe you are saying that the model is part of a discovery process, where one gets inspiration from it, tests predictions, refines the model, and so forth, bootstrapping your way up. There is no formula for how to get ideas or develop theories, and such modeling may in fact help some people. But, to paraphrase Dreyfus (1979), this approach may be something like trying to get to the moon by climbing a tree. It is impossible to examine the entire range of possible models, and thus one simply follows along within a framework defined by the initial model. And if one hits the top of the tree, one just can't go any further. One does not systematically search a space by exploring variations on a particular model. I don't see much promise that one will not always be hitting the top of the tree, because I just don't think we know enough about the representations and delineation of subsystems to be able to do the kind of detailed process modeling you want to do.

What if one parsed the processing a little bit differently? Instead of claiming there's a single module that does X, one says that it's divided up, with two modules, which are dedicated to different aspects of the problem. Well, that's going to change the whole ball game. I'm thinking again about Goodman's points about induction; there need to be many prior assumptions to generalize inductively.

RUECKL: No matter how principled the assumptions used to frame a model are, one still needs to follow the same process that I just described in terms of testing that assumption with empirical data and then refining the model to account for other sets of results.

KOSSLYN: I'm not arguing about the role of models, I'm arguing about how to interpret the results of the model when the model wasn't constructed on principled grounds, using enough information in advance. There are too many degrees of freedom; too much is up for grabs. It's a bad situation if one wants to start from scratch, messing around with networks and hoping to bootstrap one's way up.

RUECKL: I want to take issue with the notion of starting from scratch. I don't think it's typically the case that connectionist models are purely arbitrarily. Most models are motivated by knowledge about both the phenomenon of interest and the properties of connectionist networks. How reasonable that motivation is will vary across models—some will be much better motivated than others. But that's equally true for both connectionist and nonconnectionist models. Maybe the issue is this: What principles should be used to guide the construction of a model?

KOSSLYN: There are two separate kinds of considerations. First, one needs a theory of the computation, where one has an analysis of a problem that leads one to posit certain "assumptions" (some of which will correspond to the structure put into a network) that have to be made in order for an input to be mapped to a certain output. Second, one can observe behavioral dissociations, particularly following brain damage, which provide hints. But this is not enough; any given dissociation can be explained in many ways.

One implication of this perspective, then, is that one has to be extremely careful when setting up a network. First, one must be certain that one has defined the problem properly, so that the mapping the network will set out to do is a biologically plausible one. Many network modelers give only passing thought regarding the particular mapping they will examine (cf. Pinker & Prince, 1988). Second, once one has defined a well-formed problem, one must be very careful in specifying the representation of the input and the representation of the output. Unless one has good reasons for positing the properties of both, one doesn't really have anything. All one gets from the network is information about this mapping, so the details are everything.

RUECKL: I find myself in the funny position of agreeing with your premises but not your conclusions. Perhaps we disagree on the degree to which the construction of connectionist models (and probably nonconnectionist models as well) is guided by these kinds of considerations. To be sure, the rationale underlying the basic assumptions of a model aren't always made explicit in the way they are in Marr's work or, for that matter, in some of your work on high-level vision. But I'm inclined to believe that the rationale is implicit, rather than nonexistent. That is, it seems to me that the construction of connectionist models, or at least those connectionist models that are intended to be taken seriously as accounts of human behavior, are guided by assumptions about what information is used in a computation, as well as knowledge of behavioral data, including data on dissociations.

KOSSLYN: You might be right, and I'm just being greedy—as you said, research in this area is just coming into its own, and I may be asking for more clarity than is possible at this stage. In any case, I doubt that we can settle these issues at this point, and suspect that we have overlooked critical information that will help to grip on them in the future. This was fun, and I think we've sharpened the issues a little, if nothing else. I hope anybody who bothered to read this far will think about these issues and perhaps help to settle them. One person who certainly could help would be Bill Estes. I wonder, from where he sits, how he sees the field evolving. I guess we will have to ask him.

ACKNOWLEDGMENTS

Preparation of this chapter was supported by AFOSR grant 88-0012, which was supplemented by funding from the ONR, and NSF grant BNS 90-09619, awarded to the second author. This chapter was prepared by tape-recording this discussion, and then translating it into English from the transcripts; We thank Ellen Carey for transcribing the tape.

REFERENCES

Ahmad, S., & Tesauro, G. (1989). Scaling and generalization in neural networks: A case study. *Proceedings of the Annual Meeting of the Society on Natural Information Processing Systems,* 160–168.

Andersen, R. A., & Zipser, D. (1988). The role of the posterior parietal cortex in coordinate transformations for visual-motor integration. *Canadian Journal of Physiology and Pharmacology, 66,* 488–501.

Baum, E. B., & Haussler, D. (1989). What size net gives valid generalization? *Proceedings of the Annual Meeting of the Society on Natural Information Processing Systems,* 81–90.

Brousse, O., & Smolensky, P. (1989). *Virtual memories and massive generalization in connectionist combinatorial learning.* Paper presented at the 11th Annual Meeting of the Cognitive Science Society, Ann Arbor, MI.

Dreyfus, H. L. (1979). *What computers can't do.* New York: Harper Colophon.

Estes, W. K. (1975). The locus of inferential and perceptual processes in letter identification. *Journal of Experimental Psychology: General, 104,* 122–145.

Goodman, N. (1973). *Fact, fiction and forecast.* Indianapolis, IN: Bobbs-Merrill.

Hetherington, P. A., & Seidenberg, M. S. (1989). *Is there "catastrophic interference" in connectionist networks?* Paper presented at the 11th Annual Meeting of the Cognitive-Science Society, Ann Arbor, MI.

Hinton, G. E., & Sejnowski, T. J. (1986). Learning and relearning in Boltzmann machines. In D. E. Rumelhart & J. L. McClelland (Eds.), *Parallel distributed processing: Explorations in the microstructure of cognition. Volume 1: Foundations* (282–317). Cambridge, MA: MIT Press.

Hopfield, J. J. (1982). Neural networks and physical systems with emergent collective computational properties. *Proceedings of the National Academy of Sciences, USA, 81,* 3088–3092.

Kolen, J. F., & Pollack, J. B. (1990). *Scenes from exclusive-or: Back propagation is sensitive to initial conditions.* Paper presented at the 12th Annual Meeting of the Cognitive Science Society, Cambridge, MA.

Kortge, C. A. (1990). *Episodic memory in connectionist networks.* Paper presented at the 12th Annual Meeting of the Cognitive Science Society, Cambridge, MA.

Kosslyn, S. M. (1987). Seeing and imaging in the cerebral hemispheres: A computational approach. *Psychological Review*, 94:148–175.

Kosslyn, S. M., Chambris, C. F., Marsolek, C. J., and Koenig, O. (in press). Categorical versus coordinate spatial representations: Computational analyses and computer simulations. *Journal of Experimental Psychology: Human Perception and Performance.*

Lehky, S. R., & Sejnowski, T. J. (1988). Network model of shape-from shading: Neural function arises from both receptive and projective fields. *Nature, 333,* 452–454.

Marr, D. (1982). *Vision.* New York: Freeman.

Massaro, D. W. (1988). Some criticisms of connectionist models of human performance. *Journal of Memory and Language, 27,* 213–234.

McCloskey, M., & Cohen, N. J. (1989). Catastrophic interference in connectionist networks: The sequential learning problem. In G. H. Bower (Ed.), *The psychology of learning and motivation* (pp. 109–165). New York: Academic Press.

Minsky, M., & Papert, S. (1969). *Perceptrons.* Cambridge, MA: MIT Press.

O'Reilly, R. C., Kosslyn, S. M., Marsolek, C. J., & Chabris, C. F. (1990). Receptive field characteristics that allow parietal lobe neurons to encode spatial properties of visual input: A computational analysis. *Journal of Cognitive Neuroscience, 2,* 141–155.

Palmer, S. E. (1987). PDP: A new paradigm for cognitive theory. *Contemporary Psychology, 32,* 925–927.

Pinker, S., & Prince, A. (1988). On language and connectionism: Analysis of a parallel distributed processing model of language acquisition. *Cognition, 28,* 73–193.

Plaut, D., Nowlan, S., & Hinton, G. (1986). Experiments on learning by back propagation. (Tech. Report CMU-CS-86-126), Computer Science Dept., Carnegie-Mellon University, Pittsburgh, PA.

Ratcliff, R. (1990). Connectionist models of recognition memory: Constraints imposed by learning and forgetting functions. *Psychological Review, 97,* 285–308.

Rueckl, J. G. (1990). Similarity effects in word and pseudoword repetition priming. *Journal of Experimental Psychology: Learning, Memory, and Cognition, 16,* 374–391.

Rueckl, J. G., Cave, K. R., & Kosslyn, S. M. (1989). Why are "What" and "Where" processed by two cortical visual systems? A computational investigation. *Journal of Cognitive Neuroscience, 1,* 171–186.

Rueckl, J. G., Kolodny, J., & McPeek, R. M. (1990). Systematicity and structure in associative mapping functions. Manuscript submitted for publication.

Rumelhart, D., Hinton, G., & Williams, R. (1986). Learning internal representations by error propagation. In D. Rumelhart & J. McClelland (Eds.), *Parallel distributed processing: Explorations in the microstructure of cognition. Volume 1: Foundations* (318–362). Cambridge, MA: MIT Press.

Schneider, W., & Detweiler, M. (1988). A connectionist/control architecture for working memory. In G. Bower (Ed.), *The psychology of learning and motivation, Volume 21* (54–121). San Diego, CA: Academic Press.

Ungerleider, L., & Mishkin, M. (1982). Two cortical visual systems. In D. Ingle, M. A. Goodale, & R. J. W. Mansfield (Eds.), *Analysis of visual behavior.* Cambridge, MA: MIT Press.

Author Index

A

Abraham, R. H., 69, 93
Aczel, J., 49, 62, 142, 148
Ahmad, S., 255, 265
Albano, A. M., 87, 95
Alper, T. M., 53, 62
Altom, M. W., 149, 153, 158, 159, 166, 167
Anderson, J. A., 91, 93, 188, 197, 198, 207, 212, 223, 227, 245
Anderson, J. R., 236, 245
Anderson, N. H., 46, 62
Anderson, R. A., 259, 265
Andrews, J., 28, 34, 43
Ashby, F. G., 88, 89, 95, 96, 106, 113, 131, 132
Atkinson, R. C., 27, 42, 91, 93, 169, 171, 172, 173, 197, 199, 214, 223

B

Babick, A. J., 206, 225
Bachevalier, J., 235, 246
Baddeley, A. D., 204, 214, 223
Bain, J. D., 222, 224, 232, 246
Baldi, P., 229, 245
Banister, H., 46, 62
Barnes, J. M., 237, 242, 244, 245
Barnsley, M., 67, 77, 80, 93
Barsalou, L. W., 151, 166
Bartelt, H., 142, 148
Bartlett, F. C., 46, 62, 234, 245
Bartlett, R. J., 46, 62
Barto, A. G., 177, 199

Bash, D., 116, 119, 120, 132
Bashore, T. R., 87, 95
Batchelder, W. H., 91, 95
Baum, E. B., 255, 258, 265
Beltrami, E., 69, 93
Billingsly, P., 102, 112
Binder, A., 173, 174, 175, 180, 181, 184, 188, 193, 195, 197
Bjork, E. L., 120, 132
Blough, D. S., 184, 185, 186, 187, 195, 197
Blumer, A., 15, 19
Born, M., 86, 93
Borsellino, A., 207, 223
Bower, G. H., 22, 23, 29, 42, 43, 150, 153, 154, 155, 157, 158, 159, 160, 162, 164, 166, 169, 170, 171, 173, 176, 179, 180, 181, 183, 184, 188, 191, 192, 195, 197, 198, 204, 223, 236, 245
Bransford, J. D., 234, 245
Broadbent, D. E., 204, 223
Brothers, A. J., 58, 62
Brousse, O., 229, 245, 255, 265
Brown, V., 116, 119, 120, 121, 132
Burke, C. J., 2, 19, 22, 33, 43, 121, 132
Buschke, H., 205, 206, 223
Bush, R. R., 2, 19, 90, 93, 171, 173, 197

C

Campbell, J. A., 154, 155, 166, 180, 197
Campbell, N. R., 46, 62

Subject Index

Printed and bound by CPI Group (UK) Ltd, Croydon, CR0 4YY

17/10/2024

01775687-0008